The Mobile
Multimedia Business

The Mobile Multimedia Business

Requirements and Solutions

Bernd Eylert

Senior Director, Deutsche Telekom, Germany

and

Professor for M-Commerce, Wildau University, Germany

John Wiley & Sons, Ltd

Other Wiley Editorial Offices

John Wiley & Sons Inc., 111 River Street, Hoboken, NJ 07030, USA

Jossey-Bass, 989 Market Street, San Francisco, CA 94103-1741, USA

Wiley-VCH Verlag GmbH, Boschstr. 12, D-69469 Weinheim, Germany

John Wiley & Sons Australia Ltd, 42 McDougall Street, Milton, Queensland 4064, Australia

John Wiley & Sons (Asia) Pte Ltd, 2 Clementi Loop #02-01, Jin Xing Distripark, Singapore 129809

John Wiley & Sons Canada Ltd, 22 Worcester Road, Etobicoke, Ontario, Canada M9W 1L1

Wiley also publishes its books in a variety of electronic formats. Some content that appears in print may not be
available in electronic books.

Library of Congress Cataloging in Publication Data

Eylert, Bernd.
   The mobile multimedia business : requirements and solutions / Bernd Eylert.
      p.   cm.
   Includes bibliographical references and index.
   ISBN 0-470-01234-X (cloth : alk. paper)
   1. Mobile communications systems.   2. Multimedia communications.   I. Title.

   HE9713.E95 2005
   384′.09′051—dc22

                     2005012178

British Library Cataloguing in Publication Data

A catalogue record for this book is available from the British Library

ISBN-13 978-0-470-01234-5 (HB)
ISBN-10 0-470-01234-X (HB)

Typeset in 10/12pt Times by Integra Software Services Pvt. Ltd, Pondicherry, India.
Printed and bound in Great Britain by Antony Rowe Ltd, Chippenham, Wiltshire.
This book is printed on acid-free paper responsibly manufactured from sustainable forestry in which at least two trees are
planted for each one used for paper production.

Contents

Preface

Nobody can be unaware of the meteoric growth of mobile communications through to its third technology generation in only 40 years. One result is that there are now many books on mobile communications and 3G technologies, and the term 'mobile multimedia' is fast becoming a buzzword. This is hardly surprising as so many companies, universities, standards developers and consultancies are working in this fast-moving field. The forces of collaboration and competition mean that the technological advances are being widely disseminated and promoted, and this process is essential if the mobile communications industry is to progress and continue to be commercially successful.

However, as well as the many technical people involved, there is another very important group of people who are primarily concerned with 3G as a business rather than as a technology. While they are looking for a basic technical appreciation of the industry, their primary need is to understand the business aspects, especially how new markets will develop, how new services and applications will evolve, what new rules will be established to handle the regulatory implications, and the way in which these new services will impact on society in general. My objective in this book is to give this group an insight into these wider aspects of 3G.

For technicians and engineers who are developing UMTS and WLAN networks and for operators and manufacturers, I hope this book helps ensure that these wider business aspects are not forgotten as new technologies are developed. The example of supersonic transport amply illustrates how advanced technology on its own is never enough to assure commercial success.

As I seek to explain in this book, 3G has generated major new challenges for regulators and their authorities. For those who have to resolve these issues, I hope the book provides helpful background on the wider business aspects of GSM and UMTS/IMT-2000, including WLAN. I believe this to be an essential foundation for them to be able to meet the 3G regulatory challenges.

And finally, for second-year telecommunications students and postgraduates needing a basic understanding of the engineering and business essentials of mobile multimedia, I hope this book will supplement the many specialised books on GSM, UMTS and WLAN which are available.

The inspiration for the book, and much of the information I present, is a direct result of my five years work as chairman of the UMTS Forum. At its most active phase, this international and independent organisation comprised almost 300 companies and authorities drawn from across the communications industry, especially operators, manufacturers and regulators. Nearly all of the subjects in this book reflect the key issues which the Forum first identified then studied, and to which it finally proposed solutions on the basis of industry-wide consensus.

Among the more important Forum achievements were the agreement on a whole new family of technical standards in ETSI and ITU, the licensing of more than 100 networks in some 30 countries worldwide and international agreement on the allocation of new radio spectrum up to the year 2008 and beyond. Without these milestones, we would not be witnessing the evolution of 3G multimedia systems now being deployed across the world.

Through the medium of this book, I am privileged to have the opportunity to pass on some of my experiences to the generation of engineers, sales and business people who will be responsible for future mobile broadband communications. I hope that they will gain a better understanding of the wider context of the mobile multimedia business and benefit from the lessons learned by one of those who was fortunate to be involved in creating the foundation for third-generation mobile communications.

Bernd Eylert
Folkestone, UK/Münster, Germany

Foreword by Jean-Pierre Bienaimé

Just a few years ago, the reality of 'mobile data' meant little more to real-world users than sending text messages or struggling to browse share prices and weather reports via a slow WAP connection.

How things have changed! With the first wave of 3G/UMTS networks operational in Asia, Europe and the US, today's customers are experiencing their first real taste of friendly, usable multimedia on the move – from streamed news bulletins, sports highlights and music videos to high-speed Internet access on the move.

Hand in hand with the accelerating roll-out of 3G networks, the end user experience continues to improve. A growing number of attractive, affordable handsets feature dramatically increased processing power and functionality compared with their antecedents of just a few years ago. At the same time, the provisioning of new services by mobile operators is becoming steadily more in tune with the needs of real customers.

But what's coming next? The next phase of 3G/UMTS evolution, currently being standardised within The Third Generation Partnership Project (www.3gpp.org), will usher in a new era of broadband mobile. Thanks to even higher transmission speeds and closer convergence with the IP world, we will see new multimedia service possibilities . . . and the true realisation of personal Internet on the move. Increasingly, we will also come to regard UMTS as the hub of the customer's personalised mobile universe – a world of affordable access to office applications, information and entertainment services via a mesh of access methods, both wired and wireless.

When the UMTS Forum was founded in 1996, its members would have been excited at the prospect of 3G licensing on a worldwide scale, the commercialisation of more than 60 3G/UMTS networks, the availability of around 140 terminal devices and millions of customers. All this has come to pass – but not without its challenges in what has been a time of great turbulence for the mobile industry and the global economy as a whole. Looking back with hindsight, Dr Eylert's book represents a fascinating document of an industry's shared vision, and serves as a testimony to the commitment of the thousands of individuals whose involvement and passion laid the foundations for today's mobile multimedia experience – and tomorrow's. It is also a valuable record of the UMTS Forum's studies and recommendations that have contributed directly to the realisation of this vision.

Jean-Pierre Bienaimé
Chairman, UMTS Forum

Foreword by Tom Wheeler

We are as we connect. How we live our lives takes the form of the networks that allow us to interconnect. Those networks are increasingly wireless, and the new wireless reality is reversing the historical trend of commerce and culture that has existed for the past several centuries. Wireless technology is the final push away from the centralised mass society towards what Alvin Toffler has termed a 'demassified' society.

The first high-speed network connection was steam rolling on steel. We do not often think of the railroad as a communications network, yet the steam locomotive moved information farther and faster than ever before. The economic underpinnings of society changed as a result. The localised subsistence economy was replaced by centralised production, and that production was then redistributed by rail to an interconnected national market. We became as we connected: a society of mass production and mass markets moving over the fastest network ever known.

Hot on the heels of the railroad came the first electronic network, the telegraph. These messages by lightening added to the growing centralisation. Financial centres developed because stock trades could be telegraphed from distant points and executed centrally. Management centres of consolidated corporate entities developed because of the ability to control regional production activities from a central point in almost real time. Information services flowed out of these hubs to the rest of the nation, carried by electronic signals. Mass production and mass markets were joined by mass communications.

As the wired telephone followed these developments it took on the characteristics of its predecessors. 'Trunk' lines (a railroad term) carried phone calls to 'switches' (another railroad concept) that were placed in the economic centres created by the earlier networks. Rural areas did not receive telephone service until government subsidies created a countervailing force to economic centralisation. The wired telephone perpetuated the paradigm of a network pulling economic activity into centralised points.

Today there are more wireless telephone connections in the world than there are wired connections. The old centripetal force of networks, which for centuries has been drawing institutions and individuals into a centralised mass society, has been replaced by a new centrifugal network-based force that is flinging economic activity outward.

Previous network generations produced centralised physical hubs. Today's wireless network revolution is pushing activity in the opposite direction. The individual, not the place, is the new hub.

Researchers have found that the concept of 'community' has changed. Formerly 'community' meant 'physical proximity'. To today's wireless user, however, community is 'connectivity'.

Formerly we defined our 'community' by the people where we lived. Thanks to wireless technology, however, we now create our own communities whose common connection is stored in the memory of our cellphone. The new community connection exists, regardless of where individuals may be physically located.

'Alone' formerly meant 'isolated'. Now it means 'Not connected'. One day I was walking by myself through the Tuileries in Paris when my cellphone rang. A caller had dialled my Washington, DC number and had automatically been connected to me in France. I was by myself, but I certainly wasn't 'alone'.

At the root of these observations is the decentralising impact of the wireless phone that has caused 'place' to no longer exist. No more must we go to a point of presence to be in a community or to get connected. In the old network world activity took place at a centralised hub that commanded individuals, production and services. In the wireless world the user is the hub and communities form around the individual rather than the individual conforming to centralised communities.

The effects just discussed have been the result of the wireless decentralisation of voice communications. Now as wireless networks become IP networks the decentralising force is becoming even greater.

Ever since early humans painted images on cave walls information has commanded the user to come to it. Gutenberg's printing press created documents held in central repositories that commanded those in search of knowledge to come to them. The public library movement built edifices around the world to bring books to the people, but still the reader had to go to the information. Even the Internet commanded the user to come to it – first physically to huge mainframes, then by requiring individuals with PCs to go to the plug in the wall in order to access the network and the ability to go to the information.

The wireless revolution that created individual hubs for voice communications is now empowering those individual hubs with access to high-speed data capable of cutting the cord to the Internet much as the cord was cut to voice networks. For the first time in history, users are commanding information to come to them, rather than vice versa.

The network-driven centralising forces of the past several centuries are being reversed by a new network reality. The individual is increasingly taking back control from the forces of centralisation. The ultimate extension of that reality is the replacement of centralised activity hubs with individual hubs made possible by anywhere, any time wireless networks.

To promote the widest possible understanding of these profound changes in personal communication, I am delighted to commend this work by Bernd Eylert, former chairman of the UMTS Forum. From his first-hand experience of the Forum's formative years in the evolution of 3G mobile, Bernd Eylert has brought together in a single volume some of the most important developments in 3G engineering, spectrum management and regulatory principles. It highlights the new marketing concepts and commercial practices that are facilitating the deployment of the mobile multimedia business and which we are increasingly witnessing around the globe.

Tom Wheeler
Washington, DC

Acknowledgements

The Mobile Multimedia Business was a project which would not have been written without the work and help of other people, especially those involved with the UMTS Forum. I am very lucky that I had the opportunity – at least during my time as chairman – to work with so many people from different countries, companies, organisations and authorities. It would be very difficult to mention them all, but I would like to thank the following.

First, my 'daily' team: the vice-chairmen, Josef Huber of Siemens, Thomas Sidenbladh of Ericsson, and Chris Wildey of Nokia; the programme managers, Alan Hadden, formerly of One-2-One, and Steve Hearnden, formerly of BT/Cellnet; Chris Solbe, the UMTS Forum's Press and Media Relations Officer; FEI/Intellect's Director of Telecommunications Tom Wills-Sandfort and his UMTS Forum supporting team of Anne Galloway, Graham Heath, Graham MacDonald and Simon Wilson; my personal assistant Victoria Feldens of T-Mobile and all members of the Steering Group – as they were many over the years I cannot name them all.

Second, a significant part of the published results, to which this book refers, could only have been achieved through the work of many people in the UMTS Forum's different Working Groups, Task Forces and Projects represented by their Chairs as they are, inter alia:

- Market Aspects Group (Chairs Martial Guillaume of Alcatel, Olivier Burois of Lucent and Alessandro Fenyves of Itatel/Siemens);
- Technical Aspects Group and IP Multimedia Subsystem Task Force (Chair Andy Watson of Motorola/TelecomsConsultancy);
- Spectrums Aspects Group (Chairs Josef Huber of Siemens, Tim Hewitt of BT, Krister Björnsjö of Telia and Anna-Tuulia Leino of Nokia);
- Regulatory Aspects Group (Chair Thomas Sidenbladh of Ericsson);
- Information Communications Technology Group (Chair Bosco Fernandes of Siemens);
- Market Study Projects (Chair Paola Tonelli of Airtouch/Vodafone);
- WRC-2000 Spectrum Allocation Project (Chair Halina Uryga of France Telecom/Orange);
- 3GPP Co-ordination Group (Chair Antonella Napolitano of TIM);
- Task Group Naming and Addressing (Chair Gerd Grotelüschen of Mannesmann D2/Vodafone);
- Social Behaviour Project (Chair Steve Hearnden).

In addition to Chris Solbe and Paola Tonelli I would like to name others who supported me, especially with the marketing and presentation of the UMTS Forum's products to the media and the financial analysts in London, New York and Hong Kong: Eileen Healy and Terry Young of Telecompetition, Stuart Sharrock of Telemates, and Patrick Toyne Sewell, Oskar Yasar and their team at Citigate Dowe Rogerson. I would like to include the Exhibition Team with inter alia Michaela Petri-Lachnit and Elke Rose of Siemens, Rainer Wegner of BMWi and Jürg Ruprecht of Cell Wave.

My thanks also go to DWRC, namely Jane Vincent, and EITO, Carola Peter, who supported me on 'Impact on social behaviour', and Professor Barry Evans of the University of Surrey and Werner Mohr of WWRF, who supported me on '3G and beyond'.

A special thanks goes to all observers and representatives of the different international authorities, trade organisations, standardisation bodies and conference organisers with which the UMTS Forum has cooperation agreements, who shared their views with me and who supported me in their specific fields:

- ETSI and 3GPP, especially Karl-Heinz Rosenbrock, Fred Hillebrand and Adrian Scrase;
- the GSM Association, especially Michael Stocks and David Court;
- the European Commission, especially Leo Koolen, Francisco Meideros and Ruprecht Niepold.

UMTS is a trademark of ETSI (European Telecommunications Standards Institute) registered in Europe and for the benefit of ETSI members and any ETSI Standards. We have been duly authorised by ETSI to use the word UMTS, and reference to that word throughout this book should be understood as UMTS [TM].

There are many other people to thank and I apologise for not naming them all.

At this point I would like to express my thanks to T-Mobile, especially Erwin Recktenwald and Rene Obermann, for the opportunity to run the UMTS Forum as an independent chairman from 1998 to 2003, and the University of Applied Sciences at Wildau, through which I was able to develop the idea for this book during my readership in the winter term 2003/4 to students in their second year of the 'Telematik' course.

Special thanks also go to Jean-Pierre Bienaimé of France Telecom/Orange, my successor as UMTS Forum Chairman, and Tom Wheeler, former President and CEO of CTIA, a good American friend, whom I especially admire for his brilliant knowledge of our industry and his leadership skills. Both of them instantly agreed to write a foreword to this book and I am very thankful for their gracious support.

Mark Hammond, with his team at John Wiley & Sons, Ltd has been the ideal editor with his excellent support and encouragement, which has enabled me to complete this project.

I would also like to thank another old telecommunications colleague, Professor Ian Groves of King's College London, a former BT employee, who agreed to be the technical editor of this book.

When I asked Steve Foster, my good old friend (and co-airman), to support me in writing this book, he agreed immediately, although he had already guessed that it would be a tough job considering the broad spectrum of contents, the time it would take and the need to discuss structure, logic and the thread of the book as well as the corrections to be made for a non-native English speaker. He is the best person in the world that I could ask for this task and I am very grateful for his tremendous support, which brought him very close to a co-author's position (which he denies). Notwithstanding this, I am delighted to have this opportunity to highlight his brilliant work and contributions to this book.

Last, but not least, my loving thanks go to my wife Dorothee who supported me in two ways: first, she played the student's role, asking me so many questions that required greater explanation in order to make the contents clear enough for any student to understand, and second, she provided constructive criticism, sweet reason and encouragement to complete this undertaking, but above all she gave me love and moral support.

Without Steve and Dorothee I would never have been able to write a book of such complexity and volume in the given time frame.

To all of you, named or otherwise, my grateful thanks.

Folkestone, UK/Münster, Germany

1

Introduction

1.1 THE 3G MOBILE PATHFINDERS: EXISTING WIRELESS COMMUNICATION SYSTEMS

The first practical examples of mobile communications were used in many countries like the USA, the UK and Germany in military services, and played a significant role in the First World War to transfer important information from the front to headquarters to take further actions. Good and secure wireless communications were an important need for all military services – army, navy and air force. In this respect, the Second World War was a big experimental battlefield for the development and evolution of mobile radio. It was in the interests of governments that after the Second World War the military investment should be paid back by civilian use, and all western European countries started their so-called first generation of mobile communication networks.

More civilised was the introduction of maritime radio, mainly in the UK at the end of the 19th century. There was an urgent need for better communications at sea, not only for diplomacy and national security, but also in emergency cases. Certainly, the military services were in the driving seat regarding the development of wireless mobile.

So also in Germany, when it came to mobile communications on trains. Such communication was tested on longwave on the Berlin to Zossen military rail link in 1918, before Deutsche Reichspost and Deutsche Reichsbahn introduced wireless communications from trains on main lines as a regular service in 1926.

In the 1920s US police forces tested radio communication between a patrol car and headquarters, but it was still far from an operational service. Nevertheless, the lessons taken from military experiences were used for the first civilian mobile networks, IMTS in the USA, established in the late 1940s and Funktelefonsystem A in Germany established in the early 1950s.

Figure 1.1 gives an overview of the different mobile generations and some examples.

The Mobile Multimedia Business: Requirements and Solutions   Bernd Eylert
© 2005 John Wiley & Sons, Ltd

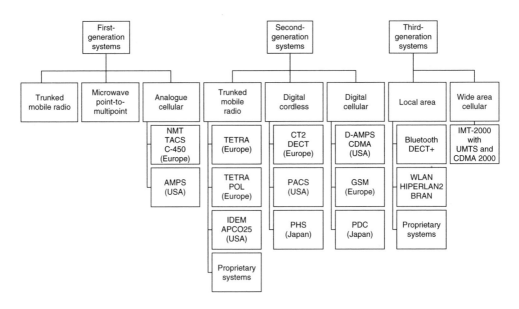

Figure 1.1 Wireless technologies. *Source*: Huber and Huber (2002).

1.1.1 ANALOGUE CELLULAR

Analogue cellular was the first mobile radio system whose popularity ensured commercial success. It was mainly used by business professionals, politicians and wealthy private individuals from the late 1940s to the early 1990s. For almost half a century it was a privileged communication system for high society. The cost per call was quite high and the equipment costs were the equivalent of a good middle class car as late as the 1970s. Because there was a lack of frequencies, which limited the number of potential users, it never became a mass-market business. For example, just after the reunification of Germany in 1990, when mobile radios were absolutely necessary and sometimes the only means of communication, the C-450 network reached a maximum of about one million users, which was much about the theoretical expected limit. Some examples are given in Tables 1.1 and 1.2.

1.1.2 CORDLESS SYSTEMS

With the introduction of GSM in Europe as the second-generation mobile system the situation changed tremendously and the use of mobile radio became widespread at all levels of society. However, there are other mobile systems in the second generation as well, which should not be ignored. First, cordless systems will be considered (see Tables 1.3 and 1.4). They gave fixed-line telephone users the new freedom of strolling around their home and garden while still being connected to the network and able to take a phone call at any part of their premises. This freedom started in America with the analogue

Table 1.1 Some examples of analogue cellular systems

Standard	Access technology	Year of introduction
NTT	FDMA	1979
NMT 450	FDMA	1981
NMT 900	FDMA	1986
APMS	FDMA	1983
C-450	FDMA	1985
ETACS	FDMA	1985
JTACS	FDMA	1988
NTACS	FDMA	1993

Source: Data adopted from T.S. Rappaport, *Wireless Communications: Principles and Practice*, Prentice Hall, Upper Saddle River, NJ, 1996.

Table 1.2 Parameters of some analogue cellular systems

Standard	Frequency band [MHz]	Channel spacing [KHz]	Number of channels
NTT	925–940/870–885	25/6.25	600/2400
	915–918.5/860–863.5	6.25	560
	922–925/867–870	6.25	480
NMT 450	453–457.5/463–467.5	25	180
NMT 900	890–915/935–960	12.5	1999
APMS	824–849/869–894	30	832
C-450	450–455.74/460–465.74	20/10	573
TACS	890–915/935–960	25	1000
ETACS	872–905/917–950	25	1240
JTACS	915–925/860–870	25/12.5	400/800
	898–901/843–846	25/12.5	120/240
	918.5–922/863.5–867	12.5	280
NTACS	915–925/860–870	25/12.5	400/800
	898–901/843–846	25/12.5	120/240
	918.5–922/863.5–867	12.5	280

Source: Data adopted from T.S. Rappaport, *Wireless Communications: Principles and Practice*, Prentice Hall, Upper Saddle River, NJ, 1996.

Table 1.3 Overview of some cordless telephone systems

Standard	Access technology	Year of introduction
CT2	FDMA	1989
DECT	TDMA	1993
PHS	TDMA	1993

Source: Data adopted from T.S. Rappaport, *Wireless Communications: Principles and Practice*, Prentice Hall, Upper Saddle River, NJ, 1996.

Table 1.4   Parameters of some cordless telephone systems

Standard	Modulation	Frequency band [MHz]	Channel spacing [KHz]	Number of channels
CT2	GFSK	864–868 (TDD)	100	40
DECT	GFSK	1880–1900 (TDD)	1728	120
PHS	π/4–DQPSK	1895–1918 (TDD)	300	77

Source: Data adopted from T.S. Rappaport, *Wireless Communications: Principles and Practice*, Prentice Hall, Upper Saddle River, NJ, 1996.

CT1 system and found a first European market with the CT1 + system. This was followed by the first digital cordless phone system, called CT2. The main driver of that technology was Motorola, which developed CT2 for the world market. However, based on European standardisation work a competitive system was established, called DECT. DECT was approved and introduced as the European cordless standard and also found its way into non-European markets, e.g. Brazil. The Personal Handy-Phone System (PHS) was developed for the Japanese market and played a similar role to CT2 and DECT in Europe. Operators ran some tests with CT2, DECT and PHS as a low tier mobile system, which customers could use at home and in dedicated areas like shopping malls and streets, railway stations, airport lounges etc. In the end this attempt failed because the extending GSM network covered all these expectations and requirements in the same way and was much cheaper. So, all cordless systems were reduced to their original territory, the home area as a wireless fixed line extension.

1.1.3 PROFESSIONAL MOBILE RADIO

PMR systems evolved before public mobile networks. Users such as police, fire, ambulance, electricity companies and transport companies (taxi, bus, coach, truck, rail) soon found that they were an essential part of their operations. The analogue systems were mostly manually controlled using central dispatchers to direct fleets in an optimum way. Later these systems became more advanced, with automatic sharing of radio channels (trunking principle), and direct access to company telephone networks and computer systems.

1.1.4 TERRESTRIAL TRUNKED RADIO (TETRA)

These systems also made the transition to digital technology, but the process has been very slow and with limited success, compared to second-generation cellular networks. In Europe, the European Telecommunications Standards Institute (ETSI) continues to define standards for the TDMA system TETRA, which competes with a more proprietary French FDMA system TETRA POL, first developed for the French police force and gendarmerie (Table 1.5). Both systems are finding a role in the market for non-public networks, but cannot be regarded as a mass-market product. Some police forces have introduced TETRA (e.g. UK, Switzerland), others TETRA POL (e.g. French police

Table 1.5 TETRA and TETRA POL

Standard	Access technology	Type	Frequency band [MHz]	Channel spacing [KHz]
TETRA	TDMA	PMR/Trunking	410–450	25/12.5
TETRA POL	TDMA	PMR/Trunking	410–450	25

force and gendamerie, FraPort in Germany). To date, most of the so-called 'Schengen' countries have not yet decided which system for security services they would like to establish for cross-border communication.

1.1.5 SHORT RANGE RADIO

A very important spin-off from the predominantly vehicle-based PMR systems was the short range network optimised for small and highly portable two-way radio terminals (Table 1.6). Such networks are used extensively in urban areas for policing, for security in shopping malls, and in industrial sites such as docks, railyards, factories, airports etc. The technical features of these systems and their coverage can be optimised to each customer's needs. However, this process is expensive and time-consuming for the customer, as individual frequencies must be licensed from the government and the radio base stations installed with all their infrastructure (masts, power, accommodation etc.). The large reductions in the costs of cellular mobiles, the increased geographical coverage of the networks and the introduction of call tariffs designed specially for large-scale users has meant that a large part of this traditional PMR market has now been taken by the public cellular networks.

PMR can therefore be said to occupy a very specific niche market which complements cellular systems. However, with the introduction of third generation the coexistence of these market segments may change again as the relationship between their different functionalities and costs alters.

Table 1.6 HIPERLAN – broadband radio access technology parameters for WLAN

Frequency band [GHz]	Operating restriction
5.150–5.350	200 mW EIRP Indoor use only Dynamic frequency selection Transmitter power control
5.470–5.650	1 W EIRP Indoor/outdoor use Dynamic frequency selection
5.650–5.725	1 W EIRP Indoor/outdoor use Dynamic frequency selection

Source: ETSI.

1.1.6 MOBILE SATELLITE SYSTEMS

It is useful to consider satellite systems according to their orbital patterns around the world. Roughly, they are classified in three groups as explained in Figure 1.2.

1.1.7 INTERNATIONAL MARITIME SATELLITE ORGANIZATION: INMARSAT

Mobile satellite systems started in the early 1980s. Again, the maritime world was the driver. The existing maritime radio systems were mainly working in the longwave, HF and VHF areas, but because of the variability and unreliability of HF wave propagation, they could not cover all maritime areas at all times, and the information carrying capacity and voice quality were poor. It was natural to study whether a satellite system could do what existing systems could not do and provide ships with more reliable emergency communications. The International Maritime Organization (IMO) developed a new system based on satellites which could identify vessels in distress any time and anywhere, and could also give them the opportunity to communicate details of their problems. Members of IMO were governmental organisations and for the communication part their Post, Telephone and Telegraph Administrations (PTT). Inmarsat is a system with four geostationary satellites connected to 50 earth stations for access to the terrestrial network. Services are offered by Inmarsat as described in Table 1.7.

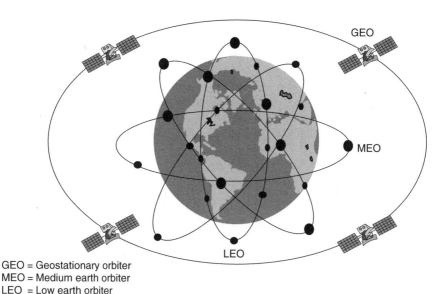

GEO = Geostationary orbiter
MEO = Medium earth orbiter
LEO = Low earth orbiter

Figure 1.2   Principles of satellite communication orbits.

Table 1.7 Inmarsat services

Service	Details
Inmarsat A, analogue service	Analogue voice: 3.1 kHz
	Fax: up to 9.6 kbps
	Data: up to 19.6 kbps
	Communication system for maritime applications
Inmarsat B/M, digital service	Voice: 16 kbps (M: 6.4 kbps)
	Fax: 9.6 kbps (M: 2.4 kbps)
	Data: up to 64 kbps (M: 2.4 kbps)
Inmarsat C	For slow data applications: 600 bps
	For fleet management
Inmarsat E	For emergency applications
Inmarsat Aero	For public communication from planes
Inmarsat P 21	10 satellites plus 2 on stand-by
	MEO with 1035 km orbit
	Two polar orbits
	Maximum capacity: 3–4 million subscribers worldwide
	Similar service as in terrestrial systems
	User data rate: 2.4 kbps

Source: Deutsche Telekom, T-Sat, 1997 and V. Jung and H.J. Warneck, *Handbuch für die Telekommunikation*, Springer-Verlag, Berlin, 1998.

1.1.8 IRIDIUM AND GLOBALSTAR

Another notable mobile satellite communication systems is IRIDIUM, a civilian variant of the 'Star Wars' communication system, an idea born in the 1980s in the USA to protect the country against nuclear threat. One unique feature of this system was the use of very large numbers of low orbit satellites with mobile stations being within the coverage area of each satellite for only a short time, thus replicating the cellular handover mechanisms seen in terrestrial cellular systems. Also unique was the use of inter-satellite links. However, political changes in Eastern Europe after the Berlin Wall came down meant that these threats receded. The American companies involved wanted to protect their investment and so they used it for civil applications. Motorola developed IRIDIUM and Qualcomm Globalstar. Both systems were built, but their businesses soon failed and came under Chapter 11 (USA bankruptcy process). IRIDIUM saw some further use, during the last Iraq war, and is still in use today. Several other satellite communication systems were planned but with little success.

IRIDIUM is a cellular mobile radio system satellite system, created by Motorola Inc. in the late 1980s, operating with 66 satellites on 6 polar LEO (780 km), with the following parameters:

- 20 gateway stations for connection to public network;
- several satellites available simultaneously;
- line-of-sight (LOS) required to satellite;

- practically no in-building communication;
- complement to terrestrial mobile radio systems;
- core network based on GSM;
- similar services as in terrestrial systems;
- user data rate 2.4 kbps.

Globalstar, an IRIDIUM competing system established by Qualcomm Inc. at the same time, operates with 48 satellites plus 8 on stand-by on 8 circular LEO with 1414 km orbit. Its technology is based on IS-95 CDMA.

Odyssey runs 12 satellites, plus several on stand-by, on 3 polar MEO with 10 335 km orbit. Finally, there is an invention of Microsoft Inc., called Teledesic. That system should run 840 satellites on LEO with 700 km orbit, specified for high data applications.

The frequency plans for some of these satellite systems are given in Table 1.8.

Under the impetus of the military and space programmes, satellite technology quickly advanced to fulfil the need for very long distance fixed communications and for broadcasting to fixed sites over very large landmasses. However, transcontinental and transoceanic fixed communications can now be delivered with greater capacity at lower cost with optical fibre. Satellite networks work as an overlay of terrestrial networks to offer communications in those areas where terrestrial coverage is impossible or too expensive. Examples are vessels on the great oceans, or for expeditions in deserts, the Artic and Antartic regions or in topographically difficult areas like high mountains. All of these mobile satellite systems use different specifications and serve different customer needs. Figure 1.3 illustrates their capacities and ranges.

1.1.9 PAGING SYSTEMS

Paging, or one-way alerting systems (bleep only or voice), first evolved for on-site systems for hospitals and similar applications. Later, large-area public paging systems were developed and have had a very different history in different areas of the world.

Table 1.8 Frequency plan for some satellite systems

Satellite system	Services links [GHz]	Feeder links [GHz]
Iridium	1.616–1.6265	29.1–29.3 uplink
		19.4–19.6 downlink
Globalstar	1.610–1.6265 uplink	5.091–5.250 uplink
	2.4835–2.500 downlink	6.875–7.055 downlink
ICO	1.985–2.150	5.150–5.250
	2.170–2.200	6.975–7.065
Ellipso	1.610–1.621	—
Asian Cellular System	1.6265–1.6605 uplink	6.425–6.725 uplink
(ACeS)	1.525–1.559 downlink	3.400–3.700 downlink

Source: Data adopted from B. Miller, Satellites free the mobile phone, *IEEE Spectrum*, **35**(3), 26–35, 1998.

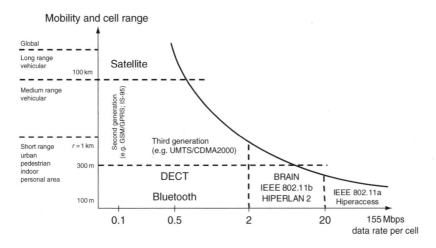

Figure 1.3 Communication systems in relation to distance.

They were first developed and introduced in the USA and South-East Asia before reaching Europe.

In Europe telephone charging operates on the principle that the calling party pays. This is different in the USA, for example, where by law both parties shared the costs of a call. That has changed in recent years, but this charging behaviour was an important factor in the introduction of paging systems. Paging allowed the caller to send a code to a receiver which is detected as an audible tone (bleep) or a few characters such as the phone number of the caller. This paging call is paid for by the sender and the called party has the freedom to respond or not, at his/her expense. Thus it is that the subsequent telephone call repeats a sharing of the costs by both parties. Later these systems evolved to carry up to several hundred characters representing a more useful message. People could also receive, e.g., weather or stock exchange information and this has proved to be an attractive service to many people, mostly working in the service industries area.

In South-East Asia the culture is different and very much based on service industries. In fact, employees are encouraged by their employers to respond by call from any distance to a much greater degree than is still the case e.g. in Europe. For that purpose paging systems were a very useful one-way information system. Technically they worked as in the USA.

In Europe, private mobile radios and cellular radio telephones were quite expensive and therefore paging systems were mostly used for quick information (one-way) to inform about meetings or changes to schedules, or for stock exchange, sports and other news (e.g. Inforuf, which started in 1991 in Germany). When mobile radios became cheaper, paging systems lost their advantage and have disappeared more and more from the European market.

However, some of the later digital European paging systems should be mentioned here. First, the Euromessage system should be addressed, which worked in Germany, France, Italy

Table 1.9 Euromessage radiopaging parameters

Standard transmission code	Transmission Rate (baud)	Channel spacing (kHz)	Code	Modulation	Mobile TX/ base TX (MHz)	Frequency hub (kHz)
CCIR Radiopaging Code No. 1 Rec. 584, POCSAG	$512 \pm 10^{-5}/1200$	20	Binary, NRZ	FSK	450–470	+4.0log. '0' −4.0log. '1'

Source: DBP Telekom FTZ 171 TR 1.

Table 1.10 Euromessage radiopaging parameters: code characteristics and organization

	Synchronisation etc.	Preamble (bit)	Number of code-word groups	Cycle time (s)	Number of slots	Timeframe of slot (s)	Recalls
Germany	POCSAG	576	24	84	3	28	No
UK	POCSAG	608	61	30	1	30	No
France	POCSAG	576	25	24	3	11.8	No
Italy	POCSAG	576	25	24	3	12	Yes

Source: DBP Telekom FTZ 171 TR 1.

and the UK based on CCIR[1] Radiopaging Code No. 1 (Recommendation 584, Geneva 1982, POCSAG-Code). Tables 1.9 and 1.10 describe the system parameters.

Another system was called ERMES (European Radio Messaging System) and was standardised by ETSI. ERMES was intended to replace Euromessage and other national radiopaging systems, but the time taken to complete the technical standards allowed cellular radio systems to satisfy the needs. So, ERMES passed into history as an idea which was overtaken by the speed of a competing and superior technology.

1.1.10 IN-FLIGHT TELEPHONE

Last but not least, another interesting system should be mentioned, although its commercial introduction in the end failed (Table 1.11, Figure 1.4). The system was called Terrestrial Flight Telephone System (TFTS). It was standardised by ETSI in the 1990s. The aim of TFTS was to have a telephone at every second place in aeroplanes flying across Europe. In the USA similar systems (InFlightphone with 500 planes, GTE Airfone with 2000 planes,

---

[1] The Comité Consultatif International de Radiocommunication (CCIR) was the predecessor of the 1992 founded International Telecommunication Union's (ITU) Radiocommunication Organisation ITU-R. Its group responsible for the standardisation of the paging systems was the Post Office Code Standarization Advisory Group (POCSAG). ITU is explained in Chapter 5.

Table 1.11   TFTS parameters

Standard transmission code	Flight level	Number of channels	Data rate (kbps)	Mobile TX/base TX (MHz)	Codec	Multiple access	Modulation
ETSI TFTS	up to FL 150	164, each up to 4 voice and 16 data channels	Synchr. 0.3–4.8 Asychr. 1.2–4.8	1.670–1.675 UL 1.800–1.805 DL	Inmarsat Aero 4–channel	TDMA	π/4 DQPSK

Sources: ETSI, DeTeMobil.

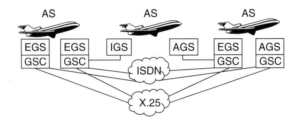

Figure 1.4   TFTS structure and components. *Sources*: ETSI, DeTeMobil.

Claircom with 1500) had been introduced at almost the same time (1992). These systems worked quite well and the operators and airlines made some money from them.

Operators and airlines expected that in-flight telephone systems, successfully introduced in the USA, could be a similar business in Europe. The market analysis conducted at that time told them that it could work successfully, but probably as a niche market. Based on the American experiences, several airlines in Europe (e.g. SAS, British Airways, Air France) introduced TFTS between 1994 and 1996, operated by their national telecommunications home carriers, who got a licence in their country. Unfortunately, the market research was found to be wrong later, when the system had started operation. One of the reasons was that flights within Europe take on average 2–3 hours only, compared to 5–6 hours in the USA. People felt there was no need to make an expensive call costing €4–5 per minute from an aeroplane when an hour or so later they could do it much cheaper from the ground through mobile or fixed-line services. Another handicap of TFTS was that it worked only when a phone call was initiated from the plane. There was no opportunity to make a call to a plane, which has much to do with safety and individuality. For security reasons airlines never tell people who is actually on the plane. Eventually, by the end of the century most TFTS equipment was de-installed from planes, because the usage was really poor and very much below expectations. This is another example of how the successful mass-market-serving GSM system killed a special dedicated communication service. Only DECT and, in limited cases, TETRA, TETRA POL and some specialised satellite systems have survived.

1.2 THE HISTORY OF 3G

The mobile communication industry has seen tremendous advances during the past 10 years, much more than ever seen before in other industrial sectors during the past 100 years. Up to the beginning of the 1990s it had been seen as a communication tool mainly for business people. That changed with the introduction of the digital communication system, especially when it came to the mass market in the second half of the 1990s. The initiative to develop a digital communication system came from Franco-German cooperation in the early 1980s, then introduced in the Conférence Européenne des Administrations des Postes et des Télé-communications (CEPT), as Group Spéciale Mobile (GSM), and with the foundation of the European Telecommunication Standards Institute (ETSI), where it developed as the Global System for Mobile Communications in today's interpretation. This political lead was supported by two other important enablers. First, the PTTs of the major European Member States signed a Memorandom of Understanding (MoU) committing to use the same technical standard and also to opening services in their capital cities by a fixed date.[2] Second, the European spectrum management authorities agreed to allocate very similar operating frequencies to the new service for all states. As well as ensuring large economies of scale in production, these two factors would make roaming throughout Europe much easier. A few years later it spread beyond the European borders to become the dominant world standard.[3]

Building on the success of GSM in Europe and especially its leadership over the USA, the European Commission decided to put in place the foundations for a similar lead in third-generation mobile communications by establishing the UMTS Task Force in 1994 through its DG XIII's RACE Project. Under the chairmanship of Bosco Fernandes of Siemens, a group of telecommunication experts[4] worked out a concept for research and development on this subject and presented to the public 'The road to UMTS – in contact anytime, anywhere, with any one' after about a year on 1st March 1996 in Brussels.

The main recommendations were published in the UMTS Task Force (1996) report and are summarised here as:

- UMTS standards must be open to global network operators and manufacturers.
- UMTS will offer a path from existing second-generation digital systems GSM900, DCS1800 and DECT.
- Basic UMTS, for broadband needs up to 2 Mbps, should be available in 2002.
- Full UMTS services and systems for mass market services in 2005.

---

[2] In the old days the PTT offices had, for political reasons, a monopoly on all these activities. That ended in the 1980s, when governments, for economic reasons, were urged to change structure and deregulate e.g. their telecommunications markets, which started in the USA with the breaking up of the old AT&T and the establishment of the so-called 'Baby Bells', and in the UK with the first private mobile radio licence given to Vodafone.

[3] Digital mobile developments in the USA were some way behind; Japan was more advanced with PDC.

[4] B. Fernandes (Chairman), E. Buitenwerf, E. Candy, D. Court, J. Desplanque, P. Dupuis, B. Eylert, A. Geiss, M.T. Hewitt, F. Hillebrand, L. Koolen, V. Kumar, J,-Y. Montfort, M. Nilsson, P. Olanders, J. Rapeli, J. da Silva, R.S. Swain, E. Vallström.

- UMTS regulatory framework (services and spectrum) must be defined by the end of 1997 to reduce the risk and uncertainties for the telecommunications industry and thereby stimulate the required investment.
- Additional spectrum (estimated $2 \times 180\,\text{MHz}$) must be made available by 2008 to allow the UMTS vision to prosper in the mass market.

Finally, according to the political mandate given by the European Parliament and Council, the Task Force recommended the establishment of the UMTS Forum as the central body charged with the elaboration of European policy towards the implementation of UMTS and based on industry-wide consensus.

Based on these recommendations, the European telecommunications industry and main administrations established the UMTS Forum at its inaugural meeting on 10th April 1996 in Bonn, Germany. Its first Chairman was Ed Candy of Orange (UK). Quite soon it was realised that because of its powerful influence and its technical commercial responsibilities, the UMTS Forum needed to be a legal entity. This was put in place on 16th December 1996 in Zurich.[5]

It could be quite useful to know that the acronym UMTS in the Task Force was defined as follows:

UMTS, the Universal Mobile Telecommunication System, will take the personal communications user into the new information society. It will deliver information, pictures and graphics direct to people and provide them with access to the next generation of information based services. It moves mobile and personal communications forward from 2nd-generation systems that are delivering mass market low-cost digital telecommunication services.

Decision No. 128/1999/EC of the European Parliament and of the Council on the coordinated introduction of a third-generation mobile and wireless communications system (UMTS) in the Community (European Commission, 1999) defines UMTS as a complex system. This definition is important because it defines the essential differences between second-generation (2G) and third-generation (3G) systems:

UMTS will be a mobile communications system that can offer significant user benefits including high-quality wireless multimedia services to a convergent network of fixed, cellular and satellite components. It will deliver information directly to users and provide them with access to new and innovative services and applications. It will offer mobile personalized communications to the mass market regardless of location, network or terminal used.

To better understand the 3G concept, the mobile communications space comprises four geographically distinct zones. These zones expand on the 2G principle of a cellular structure (Figure 1.5). In these four zones, the UMTS Forum predicted in 1999 the service environment shown in Figure 1.6.

---

[5] The UMTS Forum is an international cross-industry organisation with elected officers coming from their members (chairmen from Telia, T-Mobile and France Telecom/Orange, vice-chairmen from One2One, Siemens, Nokia and Ericsson); see www.umts-forum.org.

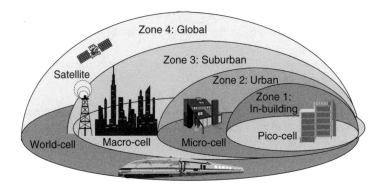

Figure 1.5 Four zone model of mobile communications. *Source*: UMTS Task Force (1996).

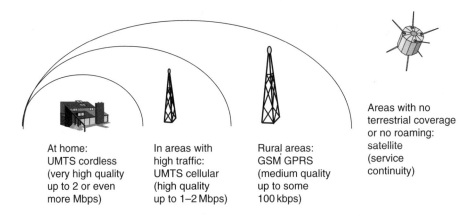

Figure 1.6 Vision of a future UMTS service environment.

In the conceptual phase of UMTS it was suggested that cordless UMTS terminals – used in the pico-zone, as explained in Figure 1.5 – be included. However, the evolution of UMTS means that this equipment is now less likely to come than it was 10 years ago and it is more likely that 'normal' UMTS handsets will cover the home environment requirements as well. Another area which may come much later than originally thought is the satellite zone. Although it seemed to be realistic in the early days, the satellite communications market has not developed as once expected (see Section 1.1.8 and Chapter 5). However, the concepts of the middle zones have proved very realistic and will be the first evolution. Indeed, these had started in Japan by autumn 2001.

Of course, while mobile communication has a high profile, it is only one component in an increasingly complex and diverse communications and information landscape. What you can see is a transformation of several industries, often referred to as 'convergence'.

One consequence is the emergence of new battlegrounds between the telecommunications, IT and media industries. In early telephone systems, there was no distinction made between the value of transporting the call and the value of the call itself. Broadcast systems, especially television, highlighted that the system which carried the signals and the content (i.e. entertainment) delivered to consumers by those signals were very different businesses. With UMTS, you can see that the media industry, specialising in the creation of all types of content will be a dominant player (Figure 1.7). These kind of developments were envisaged as long ago as 1997, in an analysis of the UMTS Forum in its first report 'A regulatory framework for UMTS' (Figure 1.8). In this figure the term 'multimedia' means mobile broadband transactions like pictures graphics, video and all kind of documents, and the term 'mobile' just means voice services. If you compare the numbers published today and those that you will see in Chapter 2 on 'Mobile communications markets', you may be surprised that the numbers given in Figure 1.8 are quite low, but the industry was always close to the ground and less excited and optimistic than would have been justified, given the reality some years later. With the tremendous success of GSM the expectations for UMTS/3G have reasonably changed.

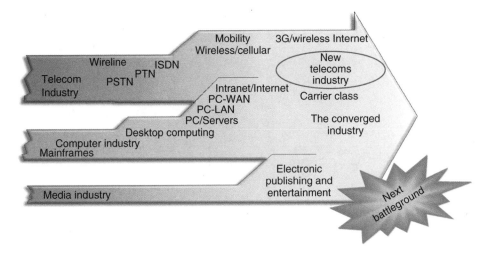

Figure 1.7   Industry transformation and convergence.

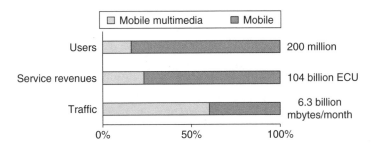

Figure 1.8   Mobile market prediction for Western Europe (EU15) by 2005. *Source*: Data from UMTS Forum (1997).

Looking at the opportunities of the media industry, Figure 1.9 illustrates the composition of media types in the UK market. This is supported by other UMTS Forum research on market opportunities (UMTS Forum, 2000a, 2001b,c, 2002a), which will be discussed in a later chapter in detail. Datacomm Research published a study in September 2003 saying that the 3G business will be driven by short video!

Multimedia services are emerging from new content markets in the so-called TIMES (= Telecommunications, Information technology (IT), Media, Entertainment, Security) services (Figure 1.10). Moving into TIMES market means extending the value chain for *all* market participants. The aim of any player is to maximise the amount of the chain which falls into their own market segment (Figure 1.11). This also means increased competition between

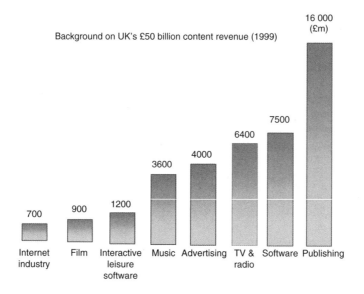

Figure 1.9   Scope for service development. *Source*: Ericsson.

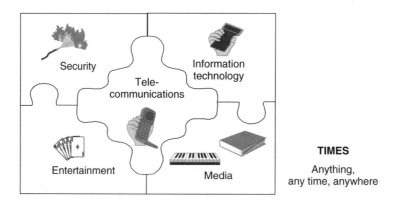

Figure 1.10   Emerged multimedia services from new convergent markets (TIMES).

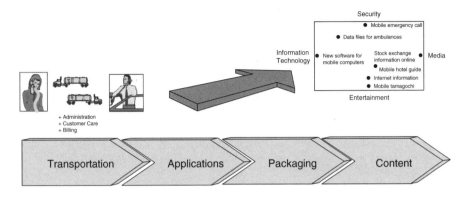

Figure 1.11 Moving into TIMES market means completion of the value chain.

more or less all market participants because traditional market shares are open to attack not only by competitors but also by, e.g., suppliers, service and/or content providers etc. It is too early to predict precisely how these new competitive forces which have been unleashed by UMTS will change the industrial landscape.

In the early days mobile multimedia was mainly seen as a technological platform and much less as a complete new system with new services. Therefore it is no surprise that the main drivers of UMTS or 3G, as it is mostly called outside Europe, were the international technical standardisation bodies like ETSI and ITU. In 1996 ETSI predicted the time scale shown in Figure 1.12 for the standardisation process for UMTS.

What kind of new mobile services can we expect from UMTS? There are many opportunities which cannot yet be explained in detail because many of them are still in the early development and trial stages. However, Figure 1.13 may give a glimpse of the new exciting world. To take

Task name	1996	1997	1998	1999	2000	2001	2002	2003	2004	2005
UMTS revised vision	▭									
Cooperative research: ACTS	▬	▬	▬							
Regulation: UMTS Forum report	▥									
Regulation: EC, ECTRA measures		▥								
Regulation: National licence conditions			▥							
Regulation: Operators identified			▥							
Operators commitment: Drafting		▨								
Operators commitment: Signature			▨							
ETSI: Basic standards studies	▩	▩								
ETSI: Freezing basic UMTS parameters		▨								
ETSI: UMTS Phase 1 standards			▩	▩						
UMTS Phase 1: System development					▬	▬				
Pre-operational trials						▬				
UMTS Phase 1: Planning, deployment						▬	▬			
UMTS Phase 1: Commercial operation										

Quelle: ETSI, MoU GSM, UMTS-Forum

Figure 1.12 UMTS time scale. *Sources*: Data from ETSI, MoU GSM and UMTS Forum.

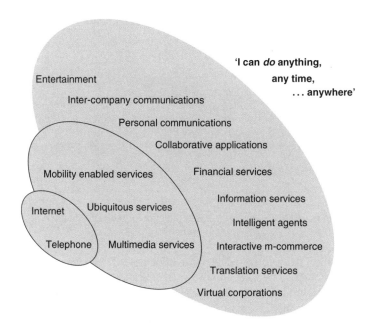

Figure 1.13   The new opportunities of mobile multimedia broadband.

just one of these service concepts, Durlacher and Lehmann Brothers have calculated what the introduction of mobile broadband m-commerce would mean in terms of ARPU (average revenue per user) (Figure 1.14).

  Globally, the three world regions – the Americas, EMEA (Europe, Middle East, Africa) and Asia – can be expected to migrate substantially to 3G type services over the 5-year period

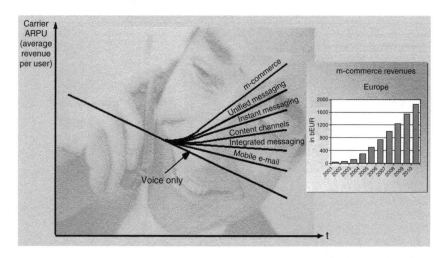

Figure 1.14   Expected m-commerce revenues for Europe. *Sources*: Durlacher report on mobile commerce (2000) and Lehmann Brothers (2000).

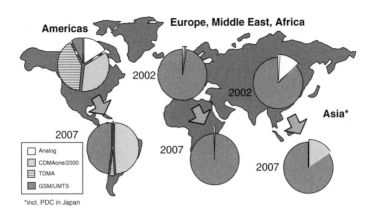

Figure 1.15 Subscriptions by regions and standards 2002–2007 (estimated). *Sources*: Data from Ovum, September 2002, and UMTS Forum.

2002–7, regardless of their choice of technology platforms, as illustrated in Figure 1.15. However, in terms of market penetration, there is no doubt that the GSM/UMTS world will be the most successful technology platform for 3G. As GSM is already dominating the second-generation mobile market, UMTS can build on this effort and offer customers seamless services regardless of the technological basis.

2

Mobile Communications Markets

This chapter introduces the reader to the importance of the market elements: the expectations of customers, customer demands, price trends, forecast revenues related to service categories, business implications, and the barriers and opportunities facing UMTS/3G.

2.1 MARKET ANALYSIS

The establishment of a new business which relies on the exploitation of new technologies requires very careful planning to ensure that the actions taken will be valid. An analysis of the economic environment, the social circumstances in which people live, and the market trends for current similar telecommunication services is recommended.

2.1.1 MARKET EXPECTATIONS AND FORECASTS AT THE TIME OF LICENSING THE FIRST 3G/UMTS NETWORKS (c. 1999)

There is no lack of information and analysis on current second-generation and planned third-generation mobile communications. Worldwide forecasts for mobile cellular subscribers, for example, are abundant and often vastly different. Definitions and methodology vary, and due to the many variables which influence consumer demand for high technology, most analysts did not attempt to forecast beyond 5 years. But even within 5 years time there are many uncertainties, e.g. the economic environment and the challenges of setting up a new business venture. Therefore, it is the author's belief that a mid-term forecast of 10 years would be more meaningful than a short term of 5 years or a long term of 15 or 20 years.

The Mobile Multimedia Business: Requirements and Solutions   Bernd Eylert
© 2005 John Wiley & Sons, Ltd

In this chapter many sources are used, mainly a series of studies submitted by Telecompetition Inc. on:

- 'The UMTS third generation market' (UMTS Forum, 2000a, 2001b,c, 2002a);
- 'Impact and opportunity: public wireless LANs and 3G business revenues' (UMTS Forum, 2002c).

They analysed over two dozen research and analyst reports, sifted through large amounts of data, and distilled the information down to those quantifiable items that impact market demand. In addition to these guiding documents, the following sources were used, inter alia:

- 'World telecommunications development report 2002' (International Telecommunication Union, 2002a);
- ABI Research, March 2004, www.abiresearch.com;
- 'European 3G users to begin surge in 2005' (Analysys Research, 2004);
- 'Mobile subscriber numbers exceed 1.5 billion' (EMC, 2004a);
- 'Mobile data to provide all revenue growth in W. Europe' (Yankee Group, 2004);
- 'Mobile subscribers to top 2B by 2006' (EMC, 2004b);
- '3G and Wi-Fi: friend or foe?' (Lovejoy, 2003);
- 'Wireless Internet users to reach 600M by '08, probe says' (RCR Wireless, 2004);
- 'T-Mobile HotSpots generate $1.4M in monthly revenue' (FierceWireless, 2004a);
- 'Vodafone struggles to build global brand in Japan' (Wall Street Journal, 2004);
- 'Mobile video: how long before it takes off?' (IEE Commentary, 2004);
- 'Camera phones face ban in courts' (CNet News and CTIA Daily News, 2004a);
- 'The Stewart report' (Independent Expert Group on Mobile Phones, 2000);
- 'Drohendes UMTS-Moratorium' (Neue Zürcher Zeitung, 2004);
- TNO (www.tno.nl);
- 3G (www.3g.co.uk);
- FierceWireless (www.fiercewireless.com);
- RCR Wireless News (www.rcrnews.com);
- CTIA Daily News (www.ctia.org);
- Silicon.com (www.silicon.com).

2.1.2 CUSTOMERS' EXPECTATIONS

Enabling any time, anywhere connectivity to the Internet is just one of the opportunities for 3G networks. 3G brings more than just mobility to the Internet. The major market opportunity builds on the unique characteristics of mobile communications to provide group messaging, location-based services, personalised information and entertainment experiences. Many new 3G services will not be Internet-based – they will be truly unique mobility services.

Data will increasingly dominate the traffic flows. Pent-up latent demand for mobile data services will jump-start 3G networks since there are more and more mobile users – over a quarter of a million every single day. By the middle of the decade, more data than voice will flow over mobile networks. This is an amazing statistic considering that mobile cellular networks today are almost exclusively voice.

Mobile subscribers will benefit from the always-on characteristic of 3G. Successful 3G service providers will navigate around the continuities that this new, always-on mobile data environment creates. Many data services are possible, and service providers with experience in marketing and billing bundled services will have an advantage.

2.1.3 THE TECHNOLOGY BACKGROUND

The success of 3G will not just come from the mere combination of two existing successful phenomena – mobility and the Internet. The real success of 3G will result from the creation of new service capabilities that genuinely fulfil a market need. Meeting market demand is not just a question of technological capability and service functionality. Creating and meeting market demand requires services and devices to be priced at acceptable levels. This requires economies of scale to be present.

The ability to benefit from economies of scale is one of the strongest market drivers for 3G services. Universal Terrestrial Radio Access (UTRA) now includes both the Direct Sequence and Time Code components of IMT-2000[1] and so embeds both the FDD access mode previously known as W-CDMA as well as the TDD modes previously known as TD-CDMA and TD-SCDMA. UTRA is now applicable to the major markets of Europe, China, South Korea and Japan (see Chapter 4). UMTS promises significant economies of scale. As UMTS is a concept, not just an access technology, the air interface will guide the economies of scales, but also the other technical components in the base stations, switches and terminals will have their part in this economic environment. However, the decision on air interface did play the most significant part.

2.1.4 MARKET DYNAMICS

Market perceptions have a greater influence on commercial success than technological realities. This should be borne in mind when considering 3G services and applications.

Although mobile Internet is only a part of the success of 3G, the enormous size of the market opportunity in mobile Internet has attracted new players from the media, financial services, entertainment, consumer electronics, fixed ISPs, as well as new 3G/mobile licensees and incumbent mobile operators.

The statistics paint a relentless picture. Internet users are growing by over half a million per day (as of 2000; Dain Rauscher Wessels). A quarter of a million people around the globe

---

[1] Study Group 11 of the ITU Telecommunication Standardisation Sector agreed five radio interface standards for the terrestrial component of IMT-2000. These are:

- IMT-DS Direct Sequence (UMTS Terrestrial Radio Access – Frequency Division Duplex (UTRA-FDD), Wideband CDMA (W-CDMA) or UMTS-FDD) using paired spectrum;
- IMT-MC Multi-Carrier (cdma2000 3X) using paired spectrum;
- IMT-TC Time Code (UTRA-TDD (Time Division Duplex) and TD-SCDMA (exact specification still to be finalised)) using unpaired spectrum;
- IMT-SC Single Carrier (UWC-136 (US TDMA)) using paired spectrum, complementary to EDGE;
- IMT-FT Frequency Time (Digital Enhanced Cordless Telecommunications (DECT)) using both paired and unpaired spectrum.

are signing up for mobile cellular service every single day (as of 2000; International Telecommunication Union, 1999). The GSM mobile cellular system reached one billion subscribers worldwide by early 2004. It took cellular just over 10 years to reach the billion mark compared with 130 years for the fixed networks (Figure 2.1).

The ITU's World Telecommunications Development Report 1999 predicted that there would be more mobile than fixed subscribers worldwide by 2007. Moreover, this view has been confirmed by ITU's 2002 Report. It shows a crossover point as early as 2002.

In contrast, this growing picture has changed much in gradient and time during the past few years. A late start of 3G has been seen, but operators now launching UMTS networks and services expect tremendous growth after 2005 and analysts such as ABI Research are supporting this view:

> More than 5 million Western Europeans will use third-generation mobile devices by the end of this year [2004], but subscriber numbers will remain low until at least 2005.

> By the end of 2004, there will be 5.3 million 3G subscribers in Western Europe, up from 600,000 at the end of 2003.

Analysys Research (2004) expects subscriber growth to stall until 2005, while network operators and handset manufacturers work through problems associated with new mass market launches. Then, from 2005 to 2009, the group expects the number of subscribers to grow a whopping 70%, to 240 million Western European subscribers. That is not much different from the picture the UMTS Forum painted in late 2000 (Figure 2.2).

However, another analyst, EMC, expects a much faster growth of the subscriber base up to 2009 (EMC, 2004a). The numbers from the Forum and from EMC have been combined in Figure 2.2. It is not known whether these numbers will mark the ceiling for the end of the decade or whether they may be underestimated. As the EMC figures are four years younger and worked out on a frequent yearly update, they may give the new business a more prosperous

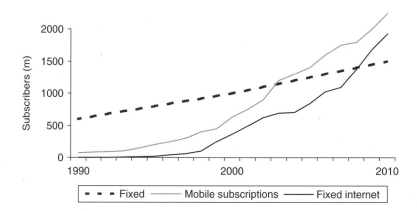

Figure 2.1   Worldwide subscribers to fixed, mobile and fixed Internet networks. *Subscribers* refers to the number of people using the service; *mobile subscriptions* refers to the number of services used. *Sources*: Data from ITU (1999) with Telecompetition Inc., July 2000.

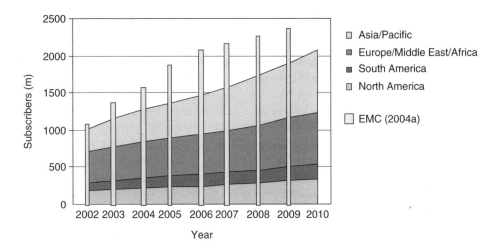

Figure 2.2   Total worldwide mobile subscriber forecast. *Sources*: UMTS Forum (2000a); EMC (2004a).

picture by about 20% compared to the Forum's numbers. In the case of 2G, forecasts have significantly underestimated actual figures. Forecasting mobile growth up to now has often been more of an art than a science.

Another good indicator for the growing market is the increasing penetration rate as suggested by Strategic Analysys in 2003 and shown in Figure 2.3.

The goal of many 3G services providers is to own as much of the subscriber's disposable income within the value chain as possible. Billing plus partnerships (content, ISP etc.) equals ownership of the subscriber. Not surprisingly, many other players have the same objective. End-users can expect to be targeted by multiple players from both ends of the value chain.

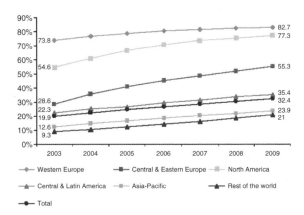

Figure 2.3   Evolution of penetration rates (in %). *Source*: Data from Analysys Research (2003).

Analyst opinions are also polarised as to who will own the customer in this new environment. The argument for the mobile services provider centres on their large subscriber base, billing systems, availability of location information, and the huge investment required to enter 3G. On the other hand, fixed Internet providers have already established partnerships with content and media providers and are, therefore, years ahead of mobile service providers in executing market leverage in those partnerships. Sometimes it seems to be difficult to separate properly between service provider and network operator, because their roles are interchangeable. For example, the virtual network operator rents capacity from the 'real' network operator but also has the relationship with the customer, so acts as service provider as well. In the following discussion, in general the term service provider is used although the role may be performed by the traditional network operator. A summary of the strengths and weaknesses of mobile and fixed Internet service providers from an analyst's viewpoint follows (Table 2.1).

2.1.5 GENERAL PRINCIPLES FOR PREPARING A MARKET STUDY

As stated at the beginning of Section 2.1, careful planning is needed before entering new markets with new technologies, and an essential tool is a comprehensive study of the market. Most operators and vendors will do that for their own strategy, but it is a very large, complex and expensive undertaking and not all have the power and capacity to do so. Therefore, if a cross-industry organisation can take over a part of that work, there are major advantages to all concerned:

- the whole industry pools its knowledge and gets a consensus view;
- small companies benefit as much as larger companies;
- administrations (regulators) and investments analysts are able to have an input into the study and derive a better picture of the new business and its challenges.

Table 2.1 Strengths and weaknesses of mobile and fixed Internet service providers

	Mobile service provider	Fixed Internet service provider
Continuity of end-user relationship	S	M
Billing system	S	W
Cost of access device (phone versus PC)	S	W
Location-based	S	W
Application simplicity	S	W
Security	S	M
Battery life	W	M
Content and commerce relationships	W	S
Network infrastructure	S	W
Corporate customer support	W	S
B2B transactions enabled	W	S

S, Strong; M, moderate; W, weak.
Sources: Data from Merrill Lynch (2000); Dain Rauscher Wessels (2000) with Telecompetition Inc. analysis, September 2000.

As stated earlier, market forecasts are mostly given for a time frame of 5 years, but as the term of a licence is more like 10–15 years, then 10 years is a more helpful horizon. This also matches more closely the period for a large new business to become firmly established. As occurred with the introduction of GSM, a delay of 1–2 years is not unusual and it reflects the complexity of a new system and its business. With mankind it takes nine months to produce a fully developed human being; with new technology it is rarely quicker!

It is important at the outset to be very clear about the scope of the studies: You must know whether you want a micro- or macroscopic approach; in other words, the details or just the big picture. Second, you must decide whether you would require a local, regional or global result. These will directly affect the complexity of the work, and therefore the time and the costs.

Last but not least, the reader must understand that forecasts are not the reality to come, but only how the market could develop if all parameters remain quite stable under the assumed circumstances.

2.1.6 STUDY METHODOLOGY

You can discuss many different methodologies to run a forecast study. One methodology may use secondary literature as a base, others may collect completely new data from customer interviews, other may take their data from companies' internal knowledge. Of course, combinations are in use as well. There are many of them and each analyst has his own favourite. When you have decided on one method, it is important to keep it as long as the study runs. Readers must know how to read the results and how to put them into the context of their business. Students should learn to know about different methodologies, to read different studies and their results; they should learn how to put the results and their interpretation into the context of their learning programme.

One methodology which has been proved successful was developed by Telecompetition, which was used for the UMTS Forum's market study. Therefore, it will be explained in considerable detail and will form the rest of this chapter.

This study used several recent secondary research reports on the subject of 3G markets as a starting point, with the goal of producing an updated market assessment for a market that is rapidly changing. Telecompetition developed the forecast contained later in this report using its proprietary *ATIVA Research Tools*®. In addition, key industry experts were interviewed to validate hypotheses and share their perspective and vision. As part of the process, Telecompetition developed a number of country-level inputs, including worldwide forecasts, propensity-to-buy and *GeoGain*™ (country weighting) scores. It has taken data based on the International Labour Organization (ILO) from about 40 countries each from Europe and Asia-Pacific, plus Canada and the USA of North America and about 80 countries from the Rest of the World.

Early in this study, the UMTS Forum identified a number of hypotheses about 3G that have been proposed by various sources. These hypotheses do not represent the predefined views of the UMTS Forum, e.g. report no. 8 (UMTS Forum, 1999b). They are a distillation of some important issues that have been raised or firm statements that have been published in the various secondary research reports. Figure 2.4 summarises the study methodology.

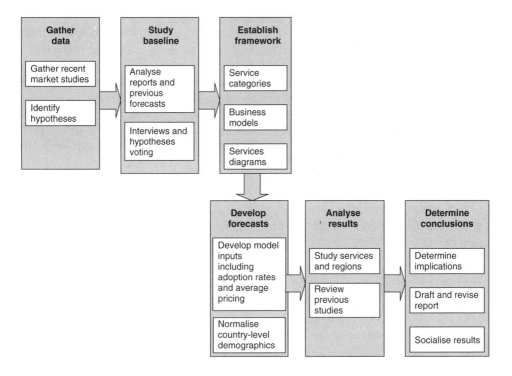

Figure 2.4   Telecompetition study methodology. *Source*: Telecompetition Inc., July 2000.

In contrast, most previous forecasts of 3G subscribers and revenues have been built around technological criteria. Subscriber number and revenue forecasts have been grouped into categories according to the data rates required, or other essentially technical criteria such as the degree of asymmetry involved in data delivery.

2.2 THE MARKET STUDIES OF THE UMTS FORUM

There are many issues and some uncertainties regarding 3G. The UMTS Forum has not attempted to re-evaluate all the issues. This has been done many times by a number of well-respected research companies. The objective of this study is to add clarity to thinking by identifying and analysing those issues critical to developing market demand, and quantifying the opportunities in light of that analysis. This has been accomplished by creating analytical frameworks for evaluating services, applications, business models and device types, and by using a structured, rigorous methodology for quantifying opportunity into regional service forecasts.

In all steps of the analysis, plausible prices and numbers have been used that reflect a likely, achievable revenue flow. This is a more conservative approach than may be taken by other analysts. The end result is a snapshot of three high-potential services with achievable 3G service provider revenue streams. The study did not attempt to quantify revenue for players other than mobile services providers, to forecast all possible services, or to recommend specific pricing strategies. The study considers demand served via 3G technologies, including satellite technology that may be used to serve remote or rural areas.

This study was produced in very close cooperation between the UMTS Forum Project Group 'Market Forecast Study' and Telecompetition. Many members of the Forum took the opportunity to work more or less intensively in that group to make the project group familiar with their own experiences. This was all discussed between members and researchers, and summarised in the delivered reports (UMTS Forum, 2000a, 2001b,c). Eventually the market study was produced in four parts: report nos 9 und 13 give the full details, report no. 17 gives an update after the continuing downturn of the economy in 2000 and 2001, and report no. 18 (UMTS Forum, 2002a) looked again at the figures after 9/11 and gives some final conclusions. The main target of the study was to calculate possible operator revenues on UMTS/3G-only for the forecast period from 2000 to 2010. Additionally, the study should give the industry and administrations a clue about the market expectations for the next decade. Furthermore, it should give operators a transparent, reasonable and serious calculation basis for the licensing bidding process.

Analysing many other existing reports of analysts and consultants and further publications of the industry, Telecompetition and the UMTS Forum chose to adopt a forecast for world-wide mobile cellular subscribers, which exceeds 30% worldwide penetration by 2010 (closely following Merrill Lynch estimates). With over 50% of the world population residing in 'low income' countries, a 30% penetration level translates into a >60% penetration level in developed countries and is therefore a reasonable estimate. In addition, estimates were made of the substitution rate of 3G over other mobile technologies. Telecompetition reviewed research by other industry experts (including Strategis, Herschel Shosteck, Merrill Lynch, Robertson Stephens and others), estimated country-level commercialisation dates, considered operator execution issues and financial limitations, and developed the 28% 3G 'share' of mobile subscribers by 2010 as shown in Figure 2.5. On a worldwide

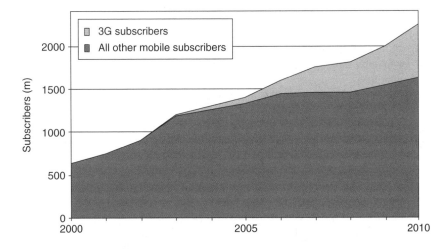

Figure 2.5   Worldwide mobile market – all subscribers. *Source*: UMTS Forum (2000a).

basis, adoption of 3G is slow in the first few years due to the time required to complete licensing and network build-out. However, in the developed countries, adoption is much faster, with Western Europe, for example, accounting for over 60% of the worldwide 3G subscribers in 2003.

One of the big benefits of the Forum's work was that the financial analysts had taken the reports as a benchmark to quote the value of 3G business in the years 2001 and 2002. With the ongoing weak economy the results of the study should be newly interpreted by the reader. For students it may be important to check other research publications on that subject and to compare the preconditions and results with those of the named study. The study that is explained now should be regarded only as a basis to understand the difficult work on market analysis, requirements and expectations, and reach some conclusions for a possible market prediction, remembering all the time that the process is not an exact science and also bearing in mind the unavoidable uncertainty which is implicit in all forecasting. The figures which are given are conservative, being set nearer to the minimum expectation rather than the maximum, which made the use of the results more serious, and resulted in an even better acceptance for financial analysts as a benchmark than other more optimistic forecasts.

2.2.1 SHORT SUMMARY OF THE STUDY

The forecasts in this study are the results of thorough analysis, using rigorous quantitative methodology with a conservative approach, to isolate the revenue that service providers can realistically retain on a service-by-service basis. They do not include the market revenue that will be attributed to other players such as wireless application service providers (WASP), content providers, device manufacturers and mobile e-commerce (m-commerce) partners.

Speculation about revenue from unknown new services has been avoided. The revenue forecast in this report represents a readily achievable goal, based on today's experiences. The result is a conservative forecast that gives 3G services providers a benchmark from which to develop future marketing strategies. The forecasts demonstrate that significant revenue flows are realistically achievable from 3G services.

Examining the entire cellular subscriber base, it is clear that 3G services can compensate for the loss of income resulting from downward pressure on voice revenues during the next decade. Coordinating the deployment of 3G networks to support roaming will accelerate this compensation effect.

This study highlights four areas of change for service providers:[2]

- business models
- industry structure
- revenue models
- market strategy.

---

[2] The term 'service provider' is introduced to indicate the changed role of the mobile operator in the 3G world.

This report presents a compelling framework for categorising and studying the majority of near-term 3G services. This framework is the basis for the structure of the analysis and forecasts. The data services are subdivided into content connectivity and mobility, then further subdivided to create the following six service categories:

- mobile Internet access
- mobile intranet/extranet access
- customised infotainment
- multimedia messaging service
- location-based service
- rich voice.

Service diagrams and business models are then added to the framework to provide a complete picture.

This study presents worldwide forecasts for 3G services that are consistent with the EU15 multimedia subscriber forecasts published in report no. 8 from the UMTS Forum (1999b). Existing forecasts of worldwide mobile subscribers and 3G substitution rates are used to determine the growth of 3G subscribers worldwide until 2010. Adoption rates are then applied to the 3G subscribers to determine worldwide service subscriptions.

These 3G service subscriptions are then allocated among 195 countries worldwide according to country-level demographics, existing mobile penetration rates, anticipated 3G commercialisation schedules, and other economic and regulatory factors. At the service and country level the forecast subscription numbers are also determined by propensity-to-buy considerations and population projections by age, occupation and industry. Addition of average pricing assumptions then allows service revenues to be forecast at a country level. The resulting country-level subscription numbers are the target market for the six 3G service categories introduced in this study.

Plausible pricing and adoption assumptions based on current pricing of analogous services existing today have been used throughout this study. This approach establishes an average price level for services based on known willingness to pay. It does not presume a pricing structure and so allows services providers flexibility in managing capacity.

Significant market potential exists for 3G services, although successful players may include only a select few in their product portfolio. These service forecasts predict a compound annual growth rate (CAGR) of over 100% growth during the forecast period, with total revenues for the forecasted services of over $320 billion[3] by 2010. This represents a cumulative revenue stream of one trillion dollars[4].

Many high-potential 3G markets will be launched by 2004, with Japan leading the way in 2001. Regional differences highlighted in the study will dictate the rate of growth in each region. Both geographic regions and economic regions are presented. Critical factors for success will be international roaming capabilities, device availability and network deployment costs.

---

[3] All financial data in this report are presented in nominal US dollars.
[4] Telecompetition Inc. estimate based on average worldwide average revenue per user (ARPU) of $35 per month per subscriber.

While this study only addresses 2000–2010 and the mere infancy of 3G, high growth is expected beyond 2010. At that time, developing countries will accelerate deployment of 3G to realise the potential for accelerating national development and closing the information gap with the developed world. The potential number of new subscriptions is staggering, with over six billion people on the planet and the opportunity for multiple service subscriptions per person. The advent of company-wide corporate services also holds promise. Historically, mobile service for business has always been sold on an individual basis. This will change with the advent of mobile intranet/extranet access and other 3G services. 3G services provider shareholder value promises to be enhanced by these developments, if the barriers and challenges identified in this report are addressed successfully.

2.2.2 STUDY FRAMEWORK

Numerous articles, reports and documents are available that discuss 3G mobile services and applications. In all this literature, there is no clear definition of the two terms. The labels 'service' and 'application' often seem to be interchangeable even within the same document. A concept such as m-commerce will be classified as a service in one report and an application in the next. Thus, the terminology serves to confuse rather than clarify.

2.2.2.1 Services and Applications

The following definitions are used throughout the book, giving a clear distinction between services and applications:

Services are the portfolio of choices offered by services providers to a user.

Example:
Services are entities that services providers may choose to charge for separately. They will be a prime differentiator between service providers in the 3G environment. Users are likely to select their preferred 3G service providers based on the options available in their product portfolio.

Different users will choose different service options. They may elect to subscribe to a personalised mobile portal offering banking facilities. They may later decide to add unified messaging. Such service options will affect the user's bill.

Applications are service enablers – deployed by service providers, manufacturers or users.

Applications are invisible to the user. They do not appear on a user's bill.

Examples:

1. A banking service would require a secure transaction application to be implemented by the service provider.
2. A unified messaging service would require voice recognition and text-to-speech applications deployed on the network or in the terminal device.
3. Individual applications will often be enablers for a wide range of services.

Under these definitions, m-commerce will most commonly be an application rather than a service. More strictly, it will be the combination of a large number of applications (e.g. security, certification, transaction recording and interchange, application execution environments etc.) that a service provider deploys to enable a range of services.

Services are entities that mobile networks deliver from the user's perspective. Applications are entities that enable the delivery of services over mobile networks. Applications are usually sourced from third-party suppliers but may also be created by mobile service providers. Some applications could be sold by service providers to the corporate market, creating a separate revenue stream, or the application cost could be bundled within the service charges.

But the main market demand is for services, not applications. Defining the universe of services used to be trivial. A simple distinction between voice services and data services was often sufficient. In the 3G world, defining a unique set of distinct categories of services is a difficult task. These difficulties reflect the wealth of opportunity opened up by 3G. They are a cause for celebration, not despair.

2.2.2.2 Service Categories

It is quite difficult to summarise all services and applications as they mushroom day by day. Therefore Telecompetition tried to categorise them as much as possible. In the UMTS Forum's market studies, six service categories were identified that are seen to represent the majority of the demand for 3G services over the next few years. The six service categories are defined from a user perspective and are intended to reflect the perception of the market. Technological distinctions have been deliberately ignored in the service definitions.[5] There is a compelling logic behind the six service categories that are illustrated in Figure 2.6.

---

[5] Technology constraints have, of course, been taken into account in the service demand forecasts.

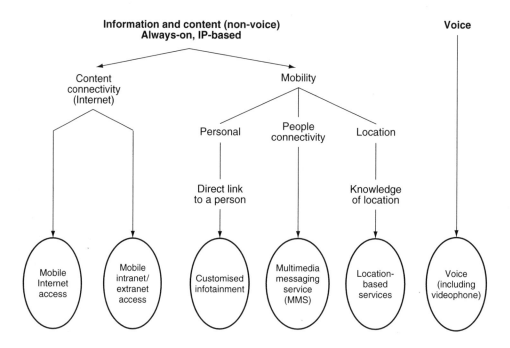

Figure 2.6   3G services framework. *Source*: Telecompetition Inc., September 2000. © UMTS Forum.

Mobile Internet and Intranet/Extranet

Rather than the voice-centric environment that has dominated the mobile world to date, 3G will be an always-on data environment. Enabling any time, any place connectivity to content on the Internet will clearly be an important role for 3G. Users will be able to add mobility to their fixed Internet experience, giving rise to what could be termed 'untethered desktop' services – mobile Internet access for the residential market segment and mobile intranet/extranet access for the business segment.

But mobility is not the only benefit provided by cellular networks. Mobile cellular networks have two distinctive features that distinguish them from fixed networks. The mobile terminal is associated with a person rather than a place, and the network knows the current location of that terminal. These are powerful features, particularly in the multimedia environment of 3G.

Customised Infotainment

Association of a terminal with a person allows the provision of a whole range of Internet-based content services tailored to the needs of the user and delivered through mobile portals. NTT DoCoMo's i-mode service is an early indicator of the potential of such customised infotainment services. Later introduced European examples on GPRS are T-Mobile's *t-zones* and Vodafone's *Vodafone-Live!*. These services based on mobile portals are a major opportunity for 3G services providers. Mobile portals encourage loyalty through the ability to personalise the selection of available content and commerce capabilities.

Multimedia messaging

Association of a terminal with a person also creates the opportunity for messaging services among closed user groups or specific communities of interest. The dramatic growth in short message service (SMS) traffic in GSM networks illustrates the demand for such messaging capabilities. The always-on characteristic of 3G networks will enable instant messaging capability, and the high data rates available will add image and video capability – near-time, in general, not real-time! – to create a multimedia messaging service (MMS).

Location-Based Services

Knowledge of the current location of a mobile terminal (which may be associated with a person or a machine) is already generating a rich portfolio of location-based services. Again, the combination of always-on connectivity and multimedia capability available with 3G adds a new dimension to this service category. Location technology not only enables specific location-based services but also enhances other service offerings, such as customised infotainment, and will be a major driver for the creation of new applications.

Rich Voice

Voice will inevitably continue to be an important service offering in the 3G environment. High data rates will allow the addition of videophone capabilities to traditional voice services. The IP environment of 3G will allow the delivery of multimedia communications within the rich voice service. Voice is the only real-time application; everything else is latency insensitive.

Conclusion

Note that high data rates, real-time and machine-to-machine communications could be a component of all six service categories.

Inevitably the boundaries between these service categories are somewhat artificial, and there is considerable overlap between the categories. Whether an individual service offering falls into one category or another could be the source of protracted (and ultimately fruitless) debate.

The service category definitions provide a framework for analysis of market demand and discussion of industry trends. They encapsulate the essential differences between the mobile and fixed environments – differences that create enormous opportunities. They incorporate the major learning lessons that have already emerged from the introduction of data services in the 2G environment.

The framework cannot, of course, include radically new service categories that have yet to be invented or implemented. Such developments will inevitably occur and will only further expand the market opportunity for 3G services.

A summary of the six service categories indicating the market segments analysed is presented in Table 2.2.

2.2.2.3 Mapping Services to Business Models

The UMTS Forum used the structure and revenue distribution of the fixed Internet industry as a baseline in developing business models for mobile Internet services providers. Figure 2.7 illustrates the functional elements of such a mobile Internet value chain. Mobile service providers have the strategic choice of participating in one or all of these areas.

Table 2.2 Services that represent the majority of the near-term 3G demand

Service name	Service description	Market segment analysed
Mobile Internet access	A 3G service that offers mobile access to full fixed Internet service provider (ISP) services with near-wireline transmission quality and functionality. Includes full Web access to the Internet as well as file transfer, email and streaming video/audio.	Consumer
Mobile intranet/ extranet access	A business 3G service that provides secure mobile access to corporate local area networks (LANs) and virtual private networks (VPNs).	Business
Customised infotainment	A consumer 3G service that provides device-independent access to personalised content anywhere, any time via structured-access mechanisms based on mobile portals.	Consumer
Multimedia messaging service	A consumer 3G service that offers near-time, multimedia messaging with always-on capabilities, allowing the provision of instant messaging. Targeted at closed user groups that can be service provider- or user-defined.	Consumer
Location-based services	A business and consumer 3G service that enables users to find other people, vehicles or machines. It also enables others to find users, as well as enabling users to identify their own location via terminal or vehicle identification.	Consumer and business
Rich voice (voice, video, and multimedia communications)	A 3G service that is real-time and two-way. It provides advanced voice capabilities (such as voice over IP (VoIP), voice-activated net access, and Web-initiated voice calls), while still offering traditional mobile voice features (such as operator services, directory assistance and roaming). As the service matures, it will include mobile videophone and multimedia communications.	Consumer and business

Sources: UMTS Forum and Telecompetition Inc. research, June 2000.

The UMTS Forum has analysed market and industry dynamics and created the following business model and market segment frameworks.

There are three probable business models for mobile service providers that will evolve with 3G mobility. The business model determines what revenue streams the mobile service provider will retain from either its partners or subscribers:

- The **mobile ISP** service provider provides ISP services only, with ISP subscription and airtime revenues.
- The **mobile portal** service provider provides a mobile portal, which includes access to selected (partner) content as well as access. Revenue sources in this model include subscription, airtime, transaction fees and advertising.
- With **mobile specialised services**, the mobile service provider provides a specialised service capability for a service-set targeted to a specific market. Revenue sources include airtime, subscription and messaging fees.

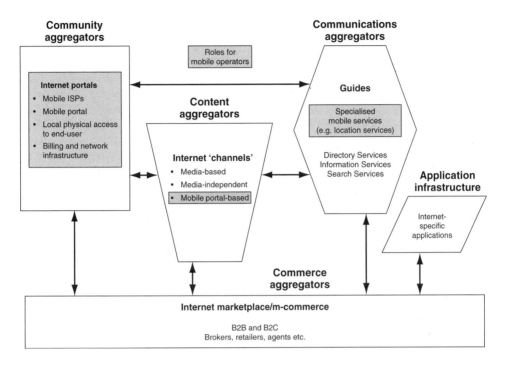

Figure 2.7 Roles for mobile service providers in the Internet value chain. *Source*: Dain Rauscher Wessels with Telecompetition Inc. analysis, January 1999.

In offering the six service categories identified in this study, the mobile service provider will use different business models, as shown in Figure 2.8. While in the real world, these three segments are not truly distinct, modelling them separately provides a cleaner picture of the market potential.

- The Internet-centric segment derives from the existing fixed Internet base and will be more likely to subscribe to a mobile computing service which provides similar functionality to that currently available from the fixed Internet. Requirements include full multimedia Internet capability and devices capable of running commonly available software applications. For modelling purposes, it was assumed that mobile service providers would provide this segment with ISP services only (i.e. that these subscribers would continue to use their fixed Internet portal even when mobile).
- The mobility-centric section derives from a mobile telephony base. Worldwide mobile subscribers will adopt 'data' capability and will be less likely (initially) to demand full Internet/Web-browsing capability and full software applications. In this segment, it was assumed that mobile service providers would provide mobile portal or specialised service functionality.
- The third segment derives from new types of services that are neither Internet- nor mobile telephony-based. These services fulfil specific user needs for mobility and content, and are typically targeted at smaller segments with specific demographic profiles. Examples include MMS targeted at teenagers and various types of location-based services.

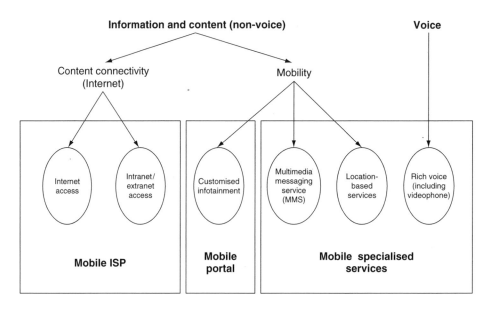

Figure 2.8 Business model framework and relationship of business models to services. *Source*: Telecompetition Inc., September 2000.

In summary, the UMTS Forum's study examines the service capabilities that UMTS can provide from a market perspective. It identifies six service categories that encompass much of the revenue potential from 3G services that can be envisaged today. It concludes that 3G is much more than the addition of mobility to the Internet. 3G has a successful future in its own right.

2.3 CALCULATIONS

2.3.1 BUSINESS MODELS FOR 3G SERVICES

Before continuing with the detailed work the reader should be introduced to a flow chart that Telecompetition created for a better understanding of the relationships between market segments, market players, services, the key data, the revenue flow and the billing procedures to get a very valid illustrative diagram for a general business model. These diagrams will also be useful for the billing processes in general, as described in Section 3.4.1.

Figure 2.9 shows a template, which Telecompetition has used to demonstrate the business models based on the service categories outlined in the market study. The service diagrams are meant as tools for continued analysis and are not intended to be complete.

In the following the reader will be introduced to each of the business models.

2.3.1.1 The Mobile Internet Access Business Model

The mobile Internet access business model targets a substantial and growing customer base of fixed Internet users who already have relationships with fixed ISPs. The Internet itself

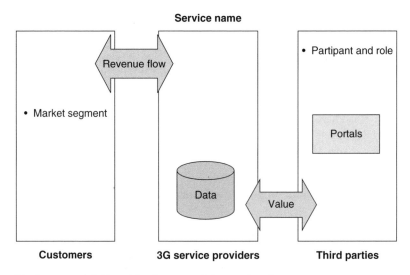

Figure 2.9 Service diagram template. © UMTS Forum.

provides the ultimate source of rich content. These customers would be willing to pay for wireless access in any other places. It will probably encompasses those mobile workers whose jobs or personal interests demand an 'anywhere' access to messages and information.

Figure 2.10 illustrates the possible business relationships among 3G service providers, customers and third parties for mobile Internet access.

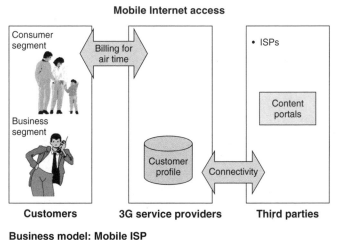

Figure 2.10 Illustrative business relationships for mobile Internet access. © UMTS Forum.

2.3.1.2 The Mobile Intranet/Extranet Access Business Model

The acknowledged demand for mobile intranet/extranet access is present in both mobile and portable scenarios. It encompasses mobile workers whose job demands any time, anywhere access to messages and information as well as management travelling outside the office. This business model may compete with those of wireless local area network (WLAN) users (see Section 2.4.1.2).

Figure 2.11 illustrates the possible business relationships among 3G services providers, customers and third parties for mobile intranet/extranet access.

2.3.1.3 The Customised Infotainment (Consumer) Business Model

The provision of customised infotainment services through a mobile portal is a viable business model. In addition to partnership arrangements with major content and media providers, possibly used to strengthen market presence through branding, mobile service providers can act as 'service bureaux'. Service bureaux offer a billing mechanism to a broad community of smaller third-party content providers who deliver content services in exchange for a share in revenue.

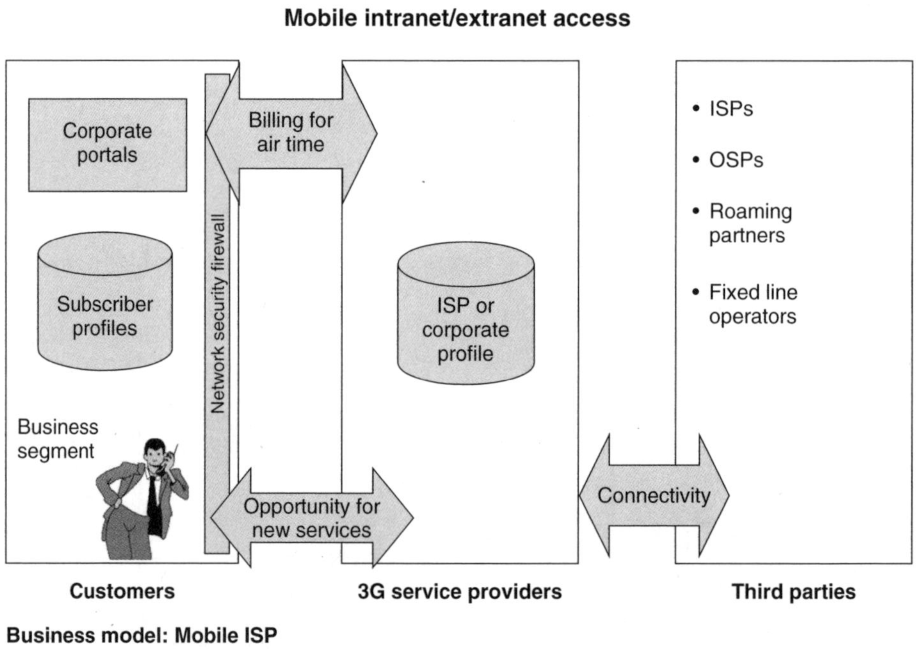

Figure 2.11 Illustrative business relationships for mobile intranet/extranet access. © UMTS Forum.

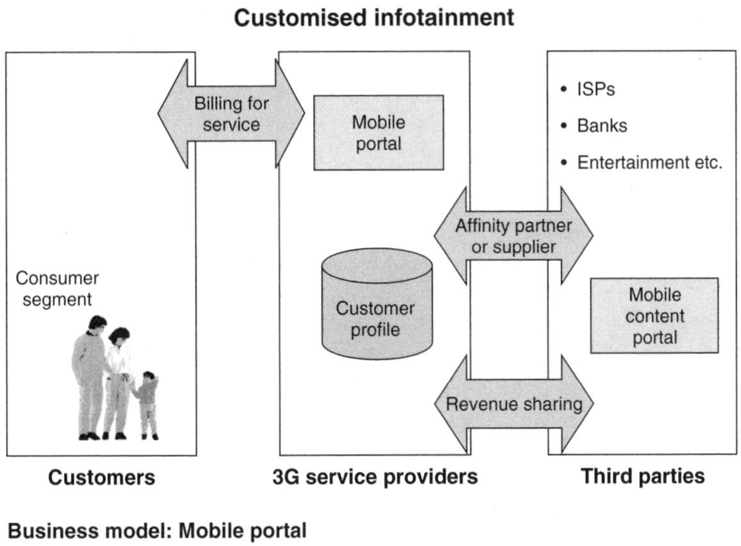

Customised infotainment

Business model: Mobile portal

Figure 2.12   Illustrative business relationships for customised infotainment. © UMTS Forum.

Figure 2.12 illustrates the possible business relationships among 3G services providers, customers and third parties for customised infotainment.

2.3.1.4 The Multimedia Messaging Service (Consumer) Business Model

Multimedia messaging should be one of the first business models as a migration of the traditional GSM SMS customer base. It is already an offer as a proven demand on today's newly deployed 3G networks.

Figure 2.13 illustrates the possible business relationships among 3G services providers, customers and third parties for the multimedia messaging service. Note that the multimedia messaging service is here only modelled for the consumer segment.

2.3.1.5 The Location-Based Services (Combined Consumer and Business) Business Model

The location-based service business model describes services that are, on one hand, inherently local, but on the other hand are not necessarily local to the user at any given time. Localised services for hotel, restaurant or taxi reservations will be required by users at both the beginning and the end of journeys. Service roaming agreements between operators are essential to the provision of really compelling location-based services.

Figure 2.14 illustrates the possible business relationships among 3G services providers, customers and third parties for location-based services.

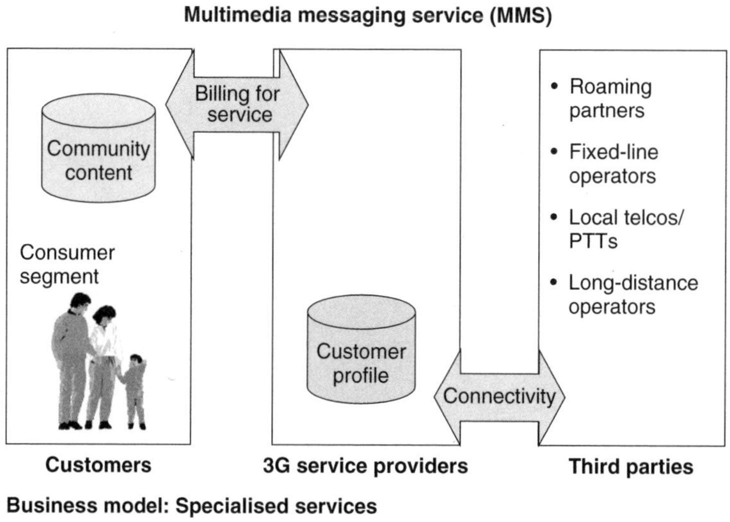

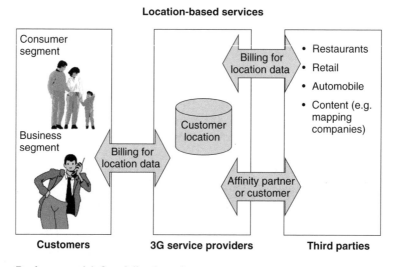

Figure 2.13   Illustrative business relationships for the multimedia messaging service. © UMTS Forum.

Figure 2.14   Illustrative business relationships for location-based services. © UMTS Forum.

2.3.1.6 The Rich Voice (Combined Consumer and Business) Business Model

The 3G rich voice service business is a real-time, two-way service for businesses and consumers that provides advanced voice capabilities such as concurrent voice and data services using Voice over IP (VoIP), voice-activated net access, and Web-initiated voice calls, while

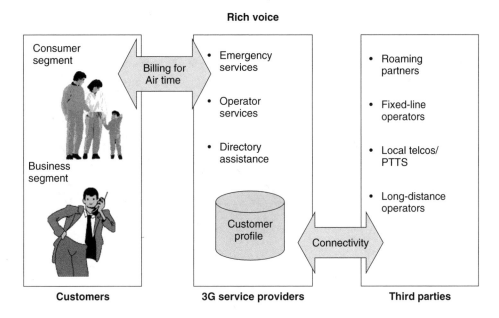

Rich voice

Business model: Specialised services

Figure 2.15 Illustrative business relationships for rich voice. © UMTS Forum.

still offering traditional mobile voice features like operator services, directory assistance and roaming. Possibly circuit-switched initially, 3G rich voice will quickly evolve to a packet-based, IP-oriented service to enable new applications.

Figure 2.15 illustrates the possible business relationships among 3G services providers, customers and third parties for rich voice.

2.3.2 USER PROFITS AND NEEDS BY SERVICE

To understand the potential target of consumer and business segments for mobile data services, it is useful to look at the characteristics and trends of existing fixed Internet users. While demographic characteristics vary widely between countries, the USA (with the highest overall number of fixed Internet users) can serve as a directional guide to understanding end-user needs. Based on data taken from some Internet research like America Online (AOL) or Boston Consulting Group (see Figures 2.16 and 2.17), the anticipated needs and requirements of users of 3G mobile services are as shown in Table 2.3.

Looking at another Internet study by the Boston Consultant Group, it is recognised for 2000 that the Internet applications in which people have the greatest interest are email and information retrieval services. Shopping and games did not play a significant role at that time. Consider whether that may change with mobile commerce.

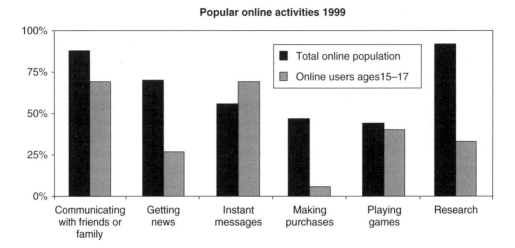

Figure 2.16 Most popular online activities by fixed Internet users – youth segment. *Source*: AOL, 'The America Online/Roper Starch Cyberstudy,' November 1999.

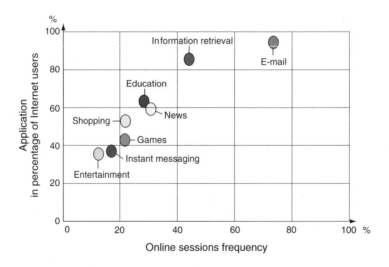

Figure 2.17 Main Internet applications 2000 (in %). *Source*: Boston Consulting Group (2000) FAZ, 30 March, p. 29.

2.3.3 PRICE TRENDS

Existing price trends in both the fixed and mobile industries are an indication of user preference in price structure. It can be expected, therefore, that user expectations for pricing will be extended into the mobile data/mobile Internet service as well. While user expectations

Table 2.3 User profiles and needs by service

Business model	Mobile ISP		Mobile portal	Mobile specialised services		
Service	Mobile Internet access	Mobile intranet/extranet access	Customised infotainment	Multimedia messaging service	Location-based services	Rich voice
Profile/demographics	Traditionally work-at-home professionals; expanding to include families and all age groups	Mobile professionals in sales, marketing, administration; increasingly distribution and logistics management	Youth and young adults up to middle age	Youth: those growing up with the Internet and young adults	Varies by service	Traditionally older and more affluent, but increasingly younger and lower income population
Needs	Anywhere, any time access to information needed	Higher productivity and greater information access to mobile workers	Need to access specific types of information (e.g. financial, recreation)	Community of interest; maintain social contact	Specific location-based information on demand	Voice access any time and anywhere
Most popular activities	News, email, research, purchases, games		Games, news, communicating, purchases	Instant messages, communicating	Purchases	Communicating

© UMTS Forum.

cannot always be reconciled with mobile service provider requirements to price and manage limited spectrum, user preferences need to be accommodated as much as possible when establishing pricing. Significant trends that mobile service providers need to consider include the following:

- A demonstrated user preference for flat rate and 'flatter rate' services. For example:
  - Mobile pricing structures are shifting from high usage-sensitive pricing to lower unit pricing with a higher subscription fee (one which includes the first 100 or so minutes).
  - In the USA, unlimited (flat-rate) Internet access is common.
- The advent of 'free' fixed Internet access.
  - This is driving up subscriber numbers and turning access from a gross profit for ISPs into a marketing cost. Adoption of broadband access is expected to reduce the negative impact on ISPs, but not enough to turn access alone into a sustainable business (UBS Warburg, 2000).
- Declining price for bandwidth.
  - In fixed data services, price per Mbps is rapidly falling. Business users quickly substitute usage sensitive (e.g. packet-based asynchronous transfer mode (ATM), frame relay) data services with flat rate (e.g. leased line) services when economic crossover is reached.
  - As wireless access is substituted for wireline access, mobile-per-minute prices approach parity with wireline pricing for both voice and data.

A direct analogy with the fixed Internet world would indicate that usage-sensitive pricing in 3G would rapidly be replaced by flat-rate plans. Market expectations for services offering 'mobile Internet access' for pricing packages will certainly be similar to those in the fixed Internet world. In 2000, Telecompetition expected that consumer surveys indicated a strong

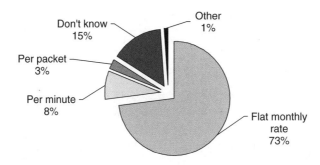

Figure 2.18 Consumer preference for flat-rate price plans. *Source*: Cahner's InStat Group, as quoted in Dain Rauscher Wessels, May 2000. © UMTS Forum.

preference for flat-rate pricing (Figure 2.18), and there is some support within the mobile industry for such developments.

2.3.3.1 Revenue Models for Services

A great deal of uncertainty exists regarding the business models to be adopted by the new industry and, therefore, the pricing structure of specific services to be offered. It should be expected that not one revenue model will dominate, but rather a variety of service-specific business and revenue models will exist. In keeping with the business model and service frameworks introduced earlier, revenue structures for the six services are defined as shown in Table 2.4.

Mobile service providers will prefer to bill for airtime, subscription and most transaction revenues to protect the relationship with their subscribers. But unlike traditional voice and SMS services, the new data services involve third-party information and content providers who will expect a share of the revenue generated.

Revenue sharing has been a feature of fixed networks for many years where intelligent networks have been used to deliver premium rate services in which the caller is charged a higher than normal rate to receive specific content. The subscriber is billed by the network service provider, which retains a proportion of the revenue to cover transport and administrative costs and passes the balance to the content provider.

Fixed network operators providing premium rate services essentially offer a billing mechanism that significantly lowers the market entry barriers for information and content providers. The number of small content providers in premium rate markets can be very large, enabling a wealth of service offerings for subscribers.

Such revenue sharing arrangements with a large number (some thousand unofficial sites) of content providers are an important component of NTT DoCoMo's i-mode service, enabling a rich variety of content to be offered to i-mode subscribers.

The premium rate industry in the fixed network environment is one of the few examples of partnerships between network service providers and content providers. It is also one of the few examples of content provision in telecommunications – and content provision will be at the heart of many 3G services.

Table 2.4 Revenue sources by service

Business model	Mobile ISP		Mobile portal	Mobile specialised services		
Service	Mobile Internet access	Mobile intranet/extranet access	Customised infotainment	Multimedia messaging service	Location-based services	Rich voice
Airtime	✓	✓	✓		✓	✓
Message				✓		
Subscription	✓	✓	✓	✓	✓	
Advertising			✓			
Transaction			✓			

© UMTS Forum.

Some lessons can be drawn from the evolution of the fixed-network premium-rate services market that could be relevant to the 3G mobile environment:

- Content providers rarely pay network operators to deliver their services. Content providers expect a revenue share.
- Competition between network operators to attract content providers soon becomes fierce. Competition focuses on attractive revenue-share arrangements for content providers, reducing the margins for network operators.
- Content providers exhibit little loyalty to network operators.
- Network operators are often the targets of bad publicity resulting from inappropriate or addictive content.
- The responsibility for handling customer complaints is often unclear. Some countries have imposed additional regulatory and consumer protection measures for content-based services.

2.3.3.2 Implications

As mobile service providers embark on new mobile data services and markets, they can expect more and different business relationships with customers and partners. Customers will expect services that meet their individual preferences at price points and quality similar to what they receive from fixed service providers. Partners will not be satisfied with a supplier relationship, but instead will expect to share in revenue opportunities. Players from all areas of the value chain will vie for 'ownership' of the customer relationship. Mobile service providers will have many new revenue opportunities, but in some targeted segments or services, they will be relegated to the role of access provider only. Brand equity will become a significant factor in establishing relationships with customers that can be leveraged in negotiating partnerships with content providers.

3G service providers can best maintain their strong market position by leveraging those capabilities unique to mobility, including location information, billing mechanisms and their existing customer base.

Table 2.5 illustrates some of the shifts required. The most significant is the overall adjustment from a static, defined industry role and customer relationship to one that is constantly evolving and requires service-specific marketing.

Table 2.5 Structural shifts in the mobile industry

	Old structure	New structure
Industry role	One role – network provider	Multiple roles – mobile portal, mobile ISP and communication infrastructure provider
Price and product	Simple price structure: airtime One product – voice	Complex, service-specific price structures: subscriptions, messages, advertising, airtime, transactions
Partnerships	Vertical integration of network infrastructure and terminal distribution	Strategic partnerships to emulate end-to-end integration

© UMTS Forum.

2.3.4 3G MARKET DEMAND FORECASTS

The demand forecast uses a most likely scenario based upon the body of secondary research, expert opinion and team analysis, using realistic price and adoption assumptions. The intent is to demonstrate likely subscription numbers and revenues given what is known today. The result is a conservative forecast that gives mobile service providers a benchmark from which to develop future marketing strategies. 'Conservative' in this context means an approach (input and output data) which is closer to the floor than to the ceiling. Forecasts for services have been derived using Telecompetition's *ATIVA Research Tools*®, a proprietary system that calculates country-level market size in four dimensions: time, product, segment and geography. The following inputs were developed for incorporation into the forecast calculations:

- industry dynamics and business models;
- worldwide 3G market size from 2000 to 2010;
- service-level propensity-to-buy scores by age, occupation and industry;
- country-level GeoGain™ scores (weighting factors) for each service that include commercialisation schedules and other significant economic and regulatory factors;
- Worldwide country-level population projections by age, occupation and industry (for 195 countries).

Telecompetition has developed comprehensive worldwide population and demographic databases using well-known sources such as International Labour Organization (ILO), US Census Bureau International Database, Rand McNally and other country-specific statistical sources. Other market factors can be considered, such as income. Uniform international demographic data for income is currently not available. These inputs are incorporated into *ATIVA Research Tools*® for calculation of country and regional forecasts. *ATIVA Research Tools*® is a registered trade mark of Telecompetition. Both, *ATIVA Research Tools*® and *GeoGain*™ are described in detail in UMTS Forum (2000a). All revenues are in nominal US dollars.

2.3.4.1 Forecast Assumptions and Inputs

In developing the demand forecasts described in later sections of this report, it was assumed that the 3G mobile industry would follow much of the business models and revenue structure of the fixed industry. Key assumptions follow:

- Different services would be offered under different business models, as described earlier. Therefore, the revenue sources for each service would also differ. For example, the mobile intranet/extranet access service uses a mobile ISP business model and would consist of subscription and airtime/access revenues only. Customised infotainment offered through a mobile portal would consist of subscription, transaction, advertising and airtime/access revenues.
- The future market position and business relationships between fixed Internet ISPs and mobile portals is highly uncertain. It is clear that both have strengths and are capable of capturing significant market share. Therefore, for modelling purposes, it was assumed that by 2010, market share is equally divided.

- Finally, the potential that exists for new mobile data services is not limited to the Internet. Services such as rich voice, multimedia messaging service, and location-based services can be offered on non-Internet platforms and new devices.

2.3.4.2 Worldwide Market Size: Subscriber and Subscription Inputs

When moving from a voice-centric to a data-centric environment, a distinction needs to be made between subscribers and subscriptions. The term 'subscribers' refers to the number of people, while the term 'subscriptions' refers to the number of services they use. In this study, the sum of the service subscriptions forecast will exceed the total number of 3G subscribers, since some subscribers will have more than one service subscription. (For example, a 3G subscriber could have both customised infotainment and multimedia messaging service.)

Worldwide service subscriptions were calculated using a top-down approach and a series of adoption curves ('S' curves) against most likely worldwide-mobile or Internet-subscriber penetration levels. To develop these adoption curves, historic substitution rates of similar products, current penetration levels, recent growth and technology constraints were analysed. Near-actual 2000 and most likely 2010 penetration levels were derived, with exponential ramp-up estimated between those two points.

2.3.4.3 Worldwide Market Size: Service Revenue Inputs

The revenue forecasts in this report are for mobile service providers' retained revenues only. They do not include the market revenue that will be attributed to other players such as content providers, device manufacturers and m-commerce partners. Therefore, many forecasts by other analysts in areas such as m-commerce may appear to indicate much larger numbers than displayed in this study. A major objective of this study has been to sift through the available data and identify realistic revenues that mobile service providers can truly expect to obtain.

For example, in the absence of generally available results from market trials to test price points for 3G services, revenue estimates were developed using current pricing of analogous services that meet similar needs and are targeted to similar markets. The underlying assumption in this approach is that users will be willing to pay at least as much for 3G services as they currently pay for a current mobile or fixed version of the capability. This approach establishes an average price level for services based on known willingness to pay. It does not presume a price structure, and so still allows for a usage-sensitive pricing scheme that would enable services providers to manage capacity by charging more bandwidth-intensive users higher prices than lighter users.

It was assumed that users will compare fixed and mobile service pricing and will make rational choices that will financially optimise their mix of mobile and fixed communication services. Therefore, it is appropriate to use fixed service prices as analogies in some cases. Service price points and usage levels were estimated using analogies as shown in Table 2.6.

Revenue sources for each forecast service were based on the three business models described earlier. For example, the customised infotainment (mobile portal) service includes revenue from airtime, subscription fees, transaction fees and advertising, while the mobile intranet/extranet access service only includes airtime and subscription fees.

Table 2.6 Analogies used for price assumptions

Service	Analogies	Year 2000[a] monthly prices per subscription
Customised infotainment (mobile portal)	i-mode	$20
Multimedia messaging service (MMS)	SMS	$8.40
Mobile intranet/extranet access	Fixed Internet access (xDSL) and subscription fees	$82 – ISP subscription and access/airtime

© UMTS Forum.
[a] These estimates were made after review of a number of secondary research sources, including Merrill Lynch, Durlacher, ARC, Robertson Stephens, Dain Rauscher Wessels, Strategis, Herschel Shosteck and others.

The multimedia messaging service includes revenue from message charges and subscription fees.

Keeping price trends and business models in mind, the following price assumptions were made in the service forecasts:

- Worldwide average prices for digital services will drop an average 16% per year for a total decline of 80% by 2010.
- The worldwide average price per message for the multimedia messaging service will drop from $0.12 to $0.035 by 2010. A premium will still be charged for higher value multimedia messages.
- Worldwide average subscription prices for mobile intranet/extranet access will decline an average 10.5% per year, for a total decline of 70% by 2010.
- Broadband access prices will drop an average 13% per year for a total decline of 80% by 2010. As a result, businesses will pay no more for mobile intranet/extranet access than they currently pay for broadband access (i.e. digital subscriber line (xDSL)) to remote locations. Subscription prices for customised infotainment will increase initially, then drop slightly over the forecast period. Increased transactions will maintain the relatively flat price trend.

Using the above assumptions, the existing price analogies and other service-specific assumptions, the average revenue per service subscription was calculated over the forecast period as shown in Figure 2.19. The steep price decline shown for the mobile intranet/extranet access service is a result of rapidly declining prices for broadband business Internet. Due to increased competition and the need to maintain mass-market price levels, average subscription revenues for customised infotainment and the multimedia messaging service increase for the first few years, then decline and flatten out over the later years in the forecast. All revenues were calculated using nominal (year 2000) US dollars.

Telecompetition propensity-to-buy scores are relative scores that weight population in each country by age, occupation and industry for each service forecast. As worldwide

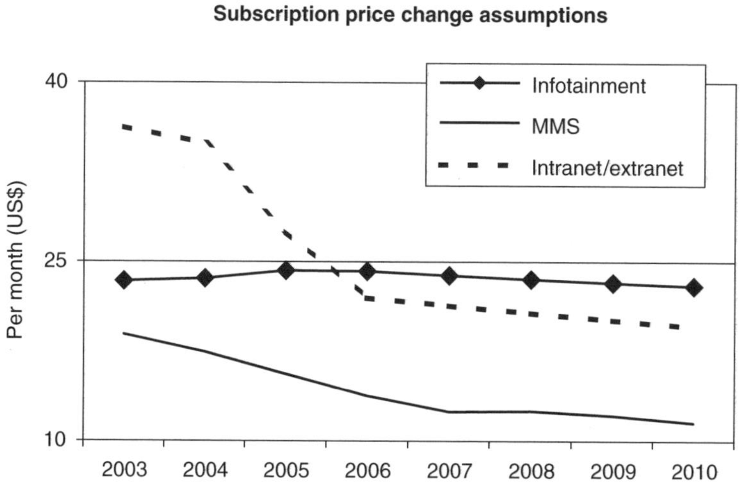

Subscription price change assumptions

Figure 2.19 Telecompetition forecast – service price assumptions. © UMTS Forum.

population data by income level is not available for all countries, propensity-to-buy by income is not considered in international forecasts. These scores are used by *ATIVA Research Tools*® to calculate the individual country-level portion of the worldwide market for each service. Service-level propensity-to-buy scores were developed through analysis of available primary and secondary market research from reputable sources. Sources reviewed include analysis by AOL, the Personal Communications Industry Association (PCIA), the Cellular Telecommunications Industry Association (CTIA) and various financial analyst reports. General propensity-to-buy assumptions are shown in Table 2.7.

2.3.4.4 Voice and Data ARPU

Before voice and data average revenue per user (ARPU) is discussed, the reader should be reminded that 3G subscribers will differ significantly from 2G subscribers: almost all 2G subscribers subscribe to the voice service. Only a few subscribe to data and fax services.

Table 2.7 Propensity-to-buy assumptions by service

Service	Propensity-to-Buy
Customised infotainment (mobile portal)	Weighted to young adults up to middle age
Multimedia messaging service (MMS)	Weighted to teenage and young adult population
Mobile intranet/extranet access	Weighted to communications-intensive industries and professional occupations

© UMTS Forum.

'Data' in 2G terms means mobile connection with a company network via IT-remote. With the introduction of GPRS it has changed a bit, so that GPRS data services also means Internet access. With 3G, subscribers will get a broad band of services to which they can subscribe individually. As it is too difficult to describe all possible services in this book, the UMTS Forum conglomerated all possible services into the six service categories described in Section 2.2.2.2. Figure 2.20 shows the subscription development for the next decade for these six service categories.

Bearing this in mind, a discussion of revenue models, used for operator revenue benchmarking, will follow.

Traditionally, ARPU is calculated by dividing total revenue by total subscribers. This can be calculated on either a worldwide basis or for individual service providers to compare revenue changes. When there is basically one product (i.e. voice) and only one subscription per subscriber, this makes sense. ARPU illustrates the net effect of declining prices and increasing usage for voice services.

ARPU is the traditional way of measuring mobile market success. In the single-service voice world, ARPU is derived by dividing total revenue by total mobile subscribers. ARPU is an adequate measure in this environment where everybody takes the same service, and where subscribers and subscriptions are the same. However, in the 3G world of various service and service bundles, multiple business models and revenue sharing, ARPU is not a meaningful measure. It does not enable service providers to measure subscriber profitability or success of individual services. Incremental revenue per subscription or subscriber becomes a more useful metric.

Generally speaking, most analysts agree that mobile data services will increase ARPU and thus reverse the current declining trend for voice-only service. For example, Merrill Lynch forecasts a 2010 total ARPU of $66 for Western Europe, almost half of which will come

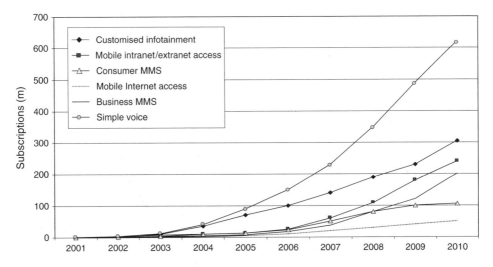

Figure 2.20 Worldwide 3G subscription – selected services (in million). © UMTS Forum (2001c).

from mobile data services. According to Merrill Lynch (2000), airtime/access for both voice and data services will continue to dominate revenue. However, non-access revenues (primarily data) includes value-added voice and data services and will increase from less than 1% to 25% of revenues by 2010. Access includes airtime generated from voice and data services (Figure 2.21).

2.3.4.5 ARPU versus Incremental Revenue per 3G Subscriber

The continued use of mobile penetration rates and ARPU as performance indicators for the future 3G environment has to be questioned. Mobile penetration rates have been a reliable measure of success for mobile markets in the past and have been considered a reasonable indicator of revenues. This may not be the case in the 3G environment, which will have characteristics closer to the Internet than to the traditional mobile world. The Internet community has also been using subscriber numbers as a virility symbol but is now learning that high subscriber numbers without an underlying revenue stream does not guarantee a sustainable long-term business.

In the 3G environment, it is important to appreciate the distinction between subscribers and subscriptions. Not all subscribers will take a subscription to all services. Service revenue forecasts include an estimate of the proportion of the subscriber base that takes a subscription to that service. Therefore, the traditional interpretation of the ARPU indicator, which is total revenue averaged across the entire subscriber base, has to be treated with caution. The traditional ARPU indicator does not accurately reflect the higher value provided by 3G subscribers over traditional voice subscribers. Therefore, this study provides both the traditional ARPU metric as well as the new incremental revenue per 3G subscriber (IRPS). This innovation undertaken by the UMTS Forum and Telecompetition has to be seen as an important step forward to compare revenues in the near future regarding multimedia services.

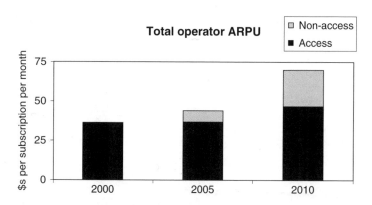

Figure 2.21 Forecasted European ARPU by access and non-access revenues. *Source*: Merrill Lynch (2000). © UMTS Forum (2001b).

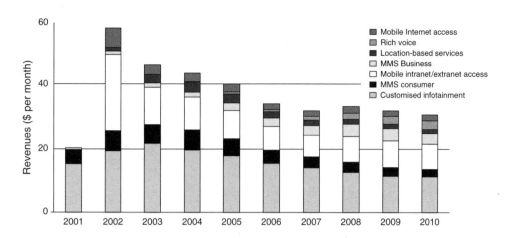

Figure 2.22 Incremental revenue per 3G subscriber by service (excluding simple voice). © UMTS Forum (2001b). Note: Due to the late start of UMTS/3G (see Section 2.4.3.1) all columns should be shifted 2–3 years later in respect of today's knowledge. Consequently this applies to all other results of this study.

In a first step Telecompetition modelled for the UMTS Forum revenue forecasts for only three 3G services. As mentioned earlier, the merit of the Telecompetition approach is that it compares ARPU with the incremental impact of the total revenue over the entire worldwide mobile subscriber base. Figure 2.22 shows the IRPS picture for all six service categories as calculated in UMTS Forum (2001b).

2.3.4.6 3G Penetration Scenario

Several analyses have been undertaken to calculate the expectable ARPU for the next decade. All analysts have in general the same view: ARPU will decline over the next few years, especially if operators continue to offer voice-only services (Figure 2.23). Most analysts share the view of the UMTS Forum that ARPU will decline from about $30 in 2001 to about $12 in 2010. You should note that this is an average worldwide number. The figures for individual countries may differ from this forecast, but have the same tendency. Under the assumption, as outlined in the UMTS Forum study, that one in three mobile users would use UMTS/3G in 2010, that would increase ARPU by 75% and a 3G penetration of 50% in 2010 would almost double the expected ARPU. That means that operators are condemned to offer mobile multimedia services if they are keen to avoid bad business or even bankruptcy in the near future. Note that these scenarios only consider additional revenue on 3G networks. Similar services may also be delivered on alternative scenarios like 2.5G or GPRS. But remember: some of the new services really need broadband 3G and they will not become successful on 2.5G, especially when subscriber numbers grow. This will become increasingly important regarding the inherent greater efficiency of 3G and the fact that it will generally be deployed over new spectrum.

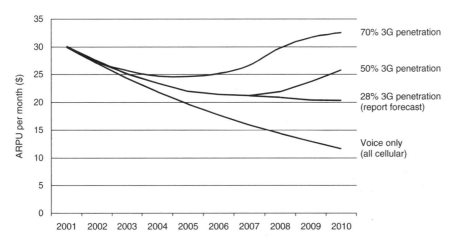

Figure 2.23   3G penetration scenario. © UMTS Forum (2001b).

In addition to the numbers and illustration given by the UMTS Forum's study, the opportunity should be taken to make the reader familiar with another view on the same subject taken by Analysys Research. It talks about revenues in general, not ARPU, but the curves show the same tendency and will give the reader a better feeling with respect to the new business (Figure 2.24).

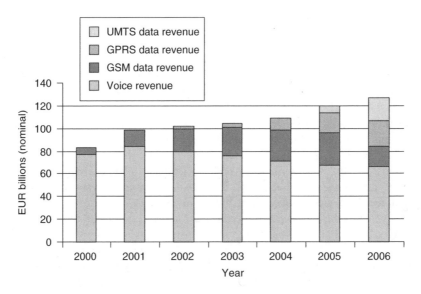

Figure 2.24   Mobile service revenue in Western Europe. *Source*: Reproduced from Analysys Research, Pricing crucial to GPRS success, 2001.

2.3.4.7 Revenue by Communication Type

Bearing in mind the remarks mentioned earlier, it is easy to understand that data will exceed voice when it comes to broadband mobility (Figure 2.25). Voice will remain an important part of the 3G service, because it is the most natural means of human communication. Anyhow, with the tremendous increase of SMS over the past few years and recognising that the younger generation (14–29 years) makes daily use of it, it will be this group that will promote data communication soon. They will drive the use of non-voice communication and, finally, will be the group which will increase data revenues over voice revenues to the benefit of the operators. Data in this context means anything that is non-voice communication.

As demonstrated in the earlier paragraph, the UMTS Forum's worldwide figures are very much in line with other research, again taken from Analysys, which, in contrast, only shows the Western European figures on that subject (Figure 2.26).

This is supported by another study, issued by Yankee Group (2004), which shows the same tendency and places even more faith in the deployment, saying that mobile voice revenues will stall in Western Europe until 2008 at around $124 billion, while data revenues will more than double over 5 years to approximately $50 billion.

2.3.4.8 Revenue by Revenue Type

Another quite interesting result of the study is the analysis of revenues by revenue type (Figure 2.27). This analysis is independent from services and service categories and looks at typical business models used today in industry. As you can see, in the Internet business, transactions and advertising play a more and more important role. If consumers do not want to pay for a service they must accept being bombarded with advertisements, which may be part of the agreement or contract they have with the service provider. On the other hand there is the classical business on airtime, subscriber acquisition and messaging. Figure 2.27 shows

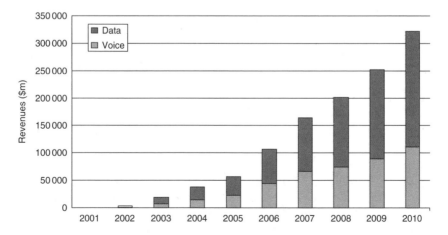

Figure 2.25   Worldwide 3G revenues – data and voice (including simple voice). © UMTS Forum (2001b).

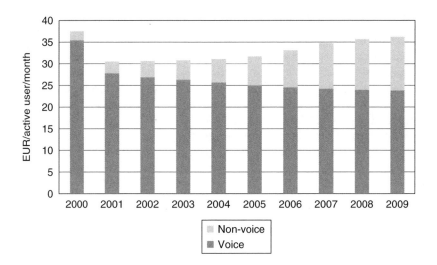

Figure 2.26 Mobile voice and non-voice ARPU in Western Europe. *Source*: Reproduced from Analysys Research, 2004.

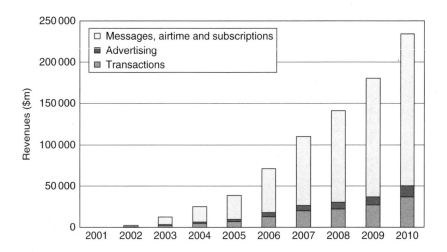

Figure 2.27 Worldwide 3G revenues by revenue type. © UMTS Forum (2000a).

the revenues operators can receive from this part of the business. The most important and surprising result is the expected revenue from advertising. It is much less than you might expect when you just look into today's TV or Internet offer. Second, subscriber acquisition and airtime is and will remain the strongest area for revenues for the next decade.

Regarding the point 'advertising versus communications', this development is proceeding and will be demonstrated by the latest Ofcom study (2004), which says that customers have spent £3.2 billion in 2003 buying access to pay TV, while advertising earned just £3.15 billion in the same time scale.

2.3.5 WORLDWIDE AND REGIONAL REVENUE FORECAST BY SERVICE CATEGORIES

2.3.5.1 Mobile Internet Access

Mobile Internet access is the access-focused counterpart to customised infotainment. It assumes that the mobile service provider is using an access-focused approach, quite similar to the ISP business models used in the fixed Internet industry. In this service category, the only revenue streams for mobile service providers consist of local mobile access (as in the fixed Internet) and the Internet access itself. The user is simply using the mobile service provider as a mechanism to 'punch through' to specific sites of interest. The model does not contain portal service provision or third-party billing or any aggregated content provision. This would be an additional field for additional revenue streams.

2.3.5.2 Intranet/Extranet

Mobile access to a business intranet or extranet is viewed as a widespread service used in combination with Internet and email by nearly all mobile, 'remote access' professionals. Therefore, it was assumed that by 2010, approximately 40% of the worldwide 3G mobile subscribers will use mobile intranet/extranet access. As the service is targeted mainly to fixed business Internet subscribers who wish to add 'mobility', the Forum assumed that service adoption will be slow until full multimedia/Web browsing with comparable quality and speed to the fixed Internet becomes available in the 2004–2005 time frame. The security issues and the general complexity of the customer base will cause the mobile intranet/extranet access service to lag consumer services by one year, delaying any significant commercial availability until 2002. That picture has changed with the economic downturn starting in 2000 and the 9/11 tragedy. As outlined in UMTS Forum (2001c, 2002a) a delay of about 2–3 years will be expected.

The revenue forecast is built upon a combination of subscription fees and airtime access. The forecast assumes that businesses will be willing to pay no more for airtime access than they currently pay for broadband remote access (i.e. DSL) and that an Internet subscription fee will also be applied. Over the forecast period, prices for airtime access decline by 70% and Internet subscription fees decline by 80%. This decline takes into account general industry trends for declining price per Mbps, discounts for negotiated contracts with businesses, and for the possibility that the mobile services provider could sell this service 'wholesale' to the business ISP or system integrator. Average monthly revenue per mobile intranet/extranet access subscription declines from $65 in 2000 to $20 by 2010.

2.3.5.3 Customised Infotainment

The demand forecast assumes customised infotainment is offered through a mobile portal. The customised infotainment revenue forecast includes subscription, airtime, advertising and transaction fees. Prices for subscriptions drop almost 50% over the 10-year horizon, while advertising fees remain constant. Transaction and airtime fees per subscription increase 46% due to increased use of the service. Average total revenue per subscription for customised infotainment rises from $20 in 2000 to $23 in 2010.

The forecast does not include subscription revenues from fixed Internet consumers who 'extend' their capability to include mobile access. This would be included in the mobile Internet access service.

2.3.5.4 Multimedia Messaging Service

Consumer

The forecast for the multimedia messaging service is restricted to a specific market segment – young consumers. The demand forecast for the multimedia messaging service uses SMS as an analogy, focusing on the teenage and young adult market and using current pricing levels as the starting price point. The forecast assumes that messages per user will increase substantially and that message prices will decrease by at least 70% by 2010. In the forecast model, the multimedia messaging service is used by about 20% of 3G subscribers, but is heavily concentrated in the teenage and young adult segments. Additional enhancements to the multimedia messaging service that could broaden the addressable market to include other age segments were not included as part of the forecast. With multimedia capability, the proportion of complex (more expensive) content and messages will increase, further increasing revenue per user.

Business

Business use of messaging services includes multimedia messaging by business professionals, messaging between machines (telemetry) and additional revenues provided by unified messaging mailboxes. Aside from professional MMS, the revenue comes from sending and receiving pictures, graphics and video clips. It is expected that the revenue will decline over the years as it may become part of the basic package for any messaging services.

2.3.5.5 Location-Based Services

The forecast of location-based services is the summation of several forecasts that consider revenue streams from consumer and business segments, both access-focused and portal-focused business models, and revenue share between 3G service providers, application service providers (ASPs) and advertisers. Most of the envisaged services are under development, which includes the revenue share between the earlier named market participants. The forecast includes just the additional revenue attributed to adding current location information to a variety of services. As there are some uncertainties, this part may be an important field to adjust over the next few years.

2.3.5.6 Voice

Simple Voice

Simple voice was calculated as an average ARPU of $30 per 3G subscriber in 2001, declining 10% per year and reaching about $12 ARPU in 2010 (see Section 2.3.4.4). It was assumed that all simple voice traffic generated by the 3G subscribers was carried over the 3G network and that all 3G subscribers would generate simple voice traffic. Scenarios on operator-specific decisions regarding load balancing between 2G, 2.5G and 3G networks were not modelled.

Rich Voice

The worldwide forecast for rich voice is a summation of real-time audio, multimedia and video services for the consumer and business segments. On the consumer side, rich voice includes revenue projections from real-time transmission of multimedia images, like digital photos, and video phone. This is distinct from consumer MMS, which is near-time, not real-time. The forecast assumes that consumer rich voice is a premium-priced service, targeted to a more affluent segment that owns a multimedia/video 3G handset/camera. As rich voice services are dependent upon higher bandwidth, which is available in 3G networks and devices, the forecast assumes that consumer rich voice services will become available around 2004–2005. Digital camera penetration was used as an analogue to determine consumer rich voice service adoption. The increased use of mobile services by business professionals and the expanded capabilities of the 3G network give 3G service providers an opportunity to offer conferencing services for both fixed and mobile users. For the business segment, revenue projections from existing worldwide conferencing services were used as an analogue.

2.3.5.7 Summary

In summary, the worldwide picture for service revenue opportunities on all six service categories in year 2000 is shown in Figure 2.28.

When the Forum calculated all these figures in early 2000 they could not have anticipated the tremendous changes resulting from the expensive licence grants in the UK and in Germany, the economic downturn that started in the second half of 2000 and then the wider turmoil affecting business which was caused by 9/11. The numbers were recalculated in late 2001 to

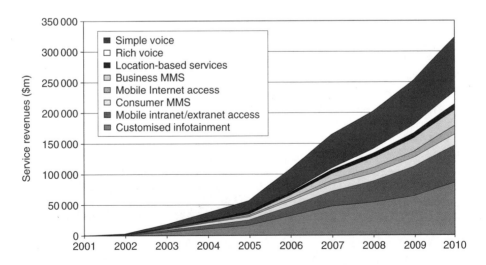

Figure 2.28 Operator revenues for 3G services (in million). © UMTS Forum (2001b).

produce a slightly different picture. On the whole, the total figures over the analysed period of 10 years did not change much: the total cumulated revenue over the decade remains at 1 trillion US dollars and the total service revenue for operators in 2010 declined from $324 billion to $322 billion, which is marginal. However, the effects of the delayed start of UMTS/3G is seen at least in the largest markets by a 2–3 year delay. In Figure 2.29, you can see the implication of the economic effects (more about the late start of UMTS/3G in Section 2.4.3.1).

2.3.6 REGIONAL FORECASTS FOR SELECTED SERVICES: 2000–2010

Regional forecasts divide the world into four regions: Europe, Asia-Pacific, North America and Rest of World (Figure 2.30). Rest of World includes primarily Latin America and Africa. Table 2.8 shows the 3G revenues listed by service categories for the administrative regions.

Figure 2.31 illustrates the differences of the data subscriber base between the four regions in the coming years.

2.3.6.1 Europe

Europe is the leading region of the world in the deployment of 3G networks. GPRS is a technological stepping-stone along the road to the emerging 3G networks. Operators installed GPRS to fill the void for the first couple of years. They started to deliver some multimedia services which do not need real broadband, but would fill the gap between decreasing ARPU on small band voice services and increasing ARPU on medium band (a few dozen kbps) data services. The forecast for countries in the European region assumes an earlier 3G commercialisation date than that assumed in other regions of the world. As you

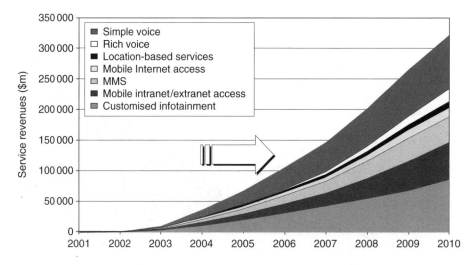

Figure 2.29   Operator revenues for 3G services after 9/11 (in million). © UMTS Forum (2001c).

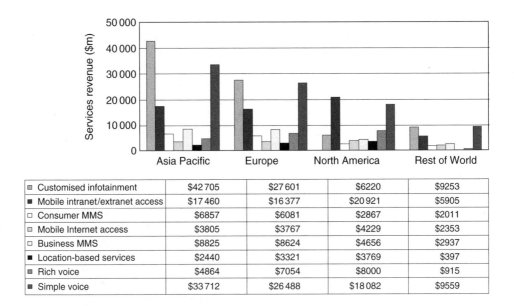

	Asia Pacific	Europe	North America	Rest of World
▨ Customised infotainment	$42 705	$27 601	$6220	$9253
▪ Mobile intranet/extranet access	$17 460	$16 377	$20 921	$5905
□ Consumer MMS	$6857	$6081	$2867	$2011
▨ Mobile Internet access	$3805	$3767	$4229	$2353
□ Business MMS	$8825	$8624	$4656	$2937
▪ Location-based services	$2440	$3321	$3769	$397
▨ Rich voice	$4864	$7054	$8000	$915
▪ Simple voice	$33 712	$26 488	$18 082	$9559

Figure 2.30 3G revenues by Region – 2010. © UMTS Forum (2001b).

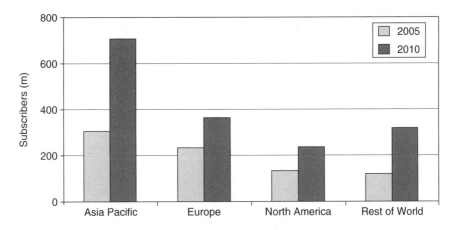

Figure 2.31 Mobile data subscribers by region, 2005 and 2010 – all technologies (in million). *Source*: Telecompetition (2002) Worldwide Mobility Report.

can see in Figure 2.30, the leading service categories will be customised infotainment followed by simple voice and mobile intranet/extranet access.

2.3.6.2 Asia-Pacific

The highest growing market in the world is Asia-Pacific. Many technologically advanced countries like Japan, Singapore, South Korea and China Hong Kong are contributing to that growth. Also, more than one third of the world's population lives in China and India. The political stability in these two countries is promoting economic growth and a customer base fertile for more mobile services. China crossed the barrier of more than 200 million mobile subscribers in 2003 and it is growing exponentially. Today, mid-2004, in China more mobile subscribers are registered than in Europe; also, the penetration rate has just passed 20% compared to more than 70% in Europe. Asia-Pacific will generate about a third of the world revenue on 3G, which is calculated to be about $118 billion in 2010. The revenue is predicted to develop in the way shown in Figure 2.32.

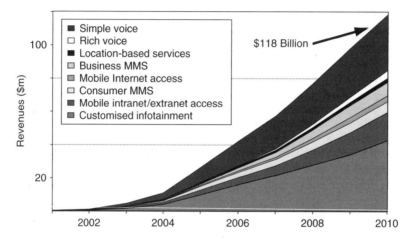

Figure 2.32 Operator revenues for 3G services in Asia-Pacific (without GPRS and CDMA2000-IX) (in $million). © UMTS Forum (2001c).

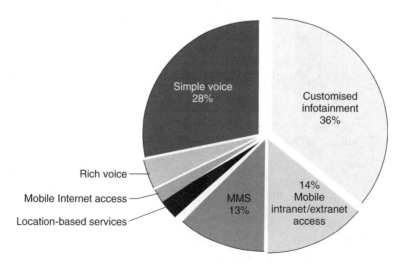

Figure 2.33 Asia-Pacific operator revenues for 3G services. © UMTS Forum (2001c).

The leading 3G service category will be customised infotainment followed by simple voice and mobile intranet/extranet access (see Figure 2.33). That sounds quite similar to Europe, but the customised infotainment is even higher, which is inter alia caused by the cultural difference and the experiences analysts could take e.g. from the Japanese market.

Although MMS is in fourth place in the service composition, Asia-Pacific is the world's leading region in the consumer MMS market, as shown in Figure 2.34.

Figure 2.35 gives the predicted numbers for all mobile data subscribers, 2/2.5G and 3G.

That China will be the leading country in the Far East may be demonstrated by the latest report of the research company Interfax (2004). Interfax predicts 3.4 million 3G users by 2005. It further predicts 38.93 million by 2007, 72.37 million by 2008, when China hosts the Olympic Games in Beijing, and 240.75 million by 2011, which implies a growth rate of over 100% per year. The only uncertainty with these figures is that China had not granted any 3G licences by mid-2004. However, rumours are that this will happen during 2004, because preparations have been in place for some two years and the granting of licences has been postponed for political and economic reasons. Another factor was that the unique Chinese version of 3G access technology, TD-SCDMA, very similar to TD-CDMA, the IMT-TC standard (see Chapter 4), was not ready and it was considered essential to give Chinese industry a fair chance to compete with its Western rivals.

2.3.6.3 North America

The North American market is different from Europe and Asia-Pacific. The USA is a couple of years behind Europe and Japan in the introduction of 3G network services. On one hand that has to do with a different spectrum policy of the FCC, which does not offer a similar amount of spectrum to operators as in Europe or Asia-Pacific. Second, the mobile business in the USA is far behind the European use of mobile phones and services. Third, the mobility industry in the USA is fragmented because of a multitude of network technologies. However, in contrast to other markets the North American customer base has a rich experience of using high speed *fixed* Internet services at very low prices. The bandwidth and capacity of 3G networks would be able to quench the thirst of this customer base for mobile Internet service.

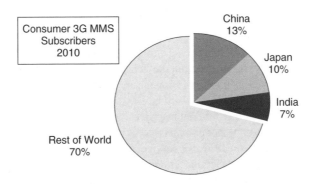

Figure 2.34   Asia-Pacific dominant in MMS consumer market by 2010. © UMTS Forum (2001c).

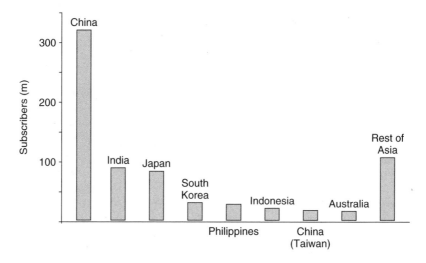

Figure 2.35 Mobile data subscribers in Asia – 2010 (in million). *Source*: Telecompetition (2002) Worldwide Mobility Report.

It is quite natural that the leading service category in that market will be mobile intranet/extranet access, followed by simple and rich voice.

2.3.6.4 Rest of the World

Last but not least, the revenue forecast picture in different parts of the Rest of the World region is very mixed. For example, there are some very rich oil producing countries in the Middle East with high GDP. On the other hand there are some very poor countries with no political stability and a very poor economic situation with little telecommunications infrastructure. It was assumed that a large portion of the countries in this region would not commercialise any 3G services until after 2007. The revenue composition in 2010 is seen as similar to that in Europe, but on a much lower level. In this study, Latin America belongs to RoW, but it will have a more significant revenue forecast.

Latin America
Latin America, especially Brazil, has been a very prosperous market for the last few years. It represents only 5% of the world market, but the general economic slowdown has not affected that market in the same way as e.g. the European market. The research data for this area is shown in Figures 2.36 and 2.37 (UMTS Forum, 2001c).

It is interesting to note that the dominant services will be simple voice, followed by MMS (consumer and business) and intranet/extranet access, shortly before customised infotainment. Except simple voice, a non-specific 3G service, the dominant 3G services are business-related services like intranet/extranet access and business MMS. This shows that the highest adoption of 3G services would be taken by the professionals, which reflects the economic

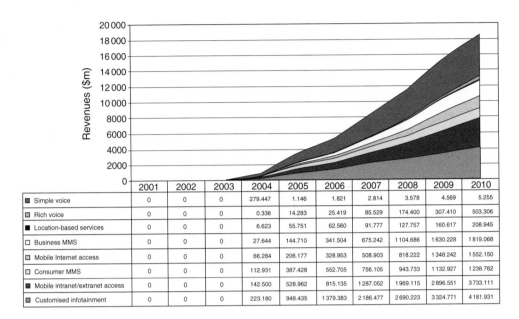

	2001	2002	2003	2004	2005	2006	2007	2008	2009	2010
■ Simple voice	0	0	0	279.447	1.146	1.821	2.814	3.578	4.569	5.255
▣ Rich voice	0	0	0	0.336	14.283	25.419	85.529	174.400	307.410	503.306
■ Location-based services	0	0	0	6.623	55.751	62.560	91.777	127.757	160.617	208.945
□ Business MMS	0	0	0	27.644	144.710	341.504	675.242	1 104.686	1 630.228	1 819.068
▣ Mobile Internet access	0	0	0	66.284	206.177	328.953	508.903	818.222	1 348.242	1 552.150
□ Consumer MMS	0	0	0	112.931	387.428	552.705	756.105	943.733	1 132.927	1 238.762
■ Mobile intranet/extranet access	0	0	0	142.500	528.962	815.135	1 287.052	1 969.115	2 896.551	3 733.111
▣ Customised infotainment	0	0	0	223.180	948.435	1 379.383	2 186.477	2 690.223	3 324.771	4 181.931

Figure 2.36 Latin America revenues – all services. *Source*: Telecompetition Inc. © UMTS Forum (2001c).

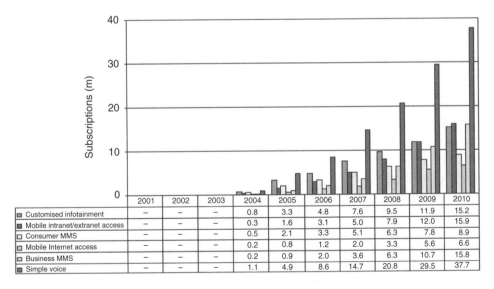

	2001	2002	2003	2004	2005	2006	2007	2008	2009	2010
▣ Customised infotainment	–	–	–	0.8	3.3	4.8	7.6	9.5	11.9	15.2
■ Mobile intranet/extranet access	–	–	–	0.3	1.6	3.1	5.0	7.9	12.0	15.9
□ Consumer MMS	–	–	–	0.5	2.1	3.3	5.1	6.3	7.8	8.9
▣ Mobile Internet access	–	–	–	0.2	0.8	1.2	2.0	3.3	5.6	6.6
□ Business MMS	–	–	–	0.2	0.9	2.0	3.6	6.3	10.7	15.8
■ Simple voice	–	–	–	1.1	4.9	8.6	14.7	20.8	29.5	37.7

Figure 2.37 Latin American subscriptions – selected services. *Source*: Telecompetition Inc. (2001). © UMTS Forum.

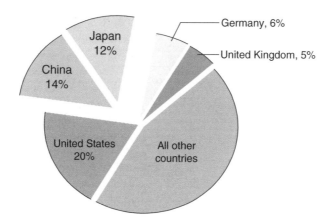

Figure 2.38   Top five countries for 3G services in 2010. © UMTS Forum 2001 and Telecompetition, August 2000.

situation in the Latin American market very well. However, as in all other regions, customised infotainment plays a significant role, reflecting the consumer wishes (not needs) in the threshold and developing countries.

2.3.6.5 Top Five Countries for 3G services in 2010

Before this section of the study finishes, the conclusions regarding the top five countries for 3G services in the world should be summarised: the USA would lead because it is expected that the US market will increase tremendously over the next few years. As the mobile penetration rate in the USA in 2000 was about 30%, the evolution of that market would soon be driven by the GSM community and by the end of the decade it will reach a similar penetration rate as in Europe, which will lead to a 20% share of 3G services in 2010. Although China is the country with the highest population, it will cover only 14% of the worldwide services in 2010, followed by Japan, which has just a sixth of the Chinese population. The leading country on 3G services in 2010 in Europe is expected to be Germany, closely followed by the UK. Summarising these facts in Figure 2.38, it is evident that three in five services in 2010 will have their market in the top five countries.

2.4 EXPECTATIONS

2.4.1 BUSINESS IMPLICATIONS

In this section some business issues that might impact the 3G service forecasts will be discussed. But before doing so, the reader should be introduced to a discussion between the leading 2G operators through their organisation, the GSM MoU, which was undertaken in the mid-1990s to get some ideas about the direction in which the mobile world, and especially the 3G world, would develop. The GSM MoU established a so-called 3GIG and asked this group to develop a vision on 3G from the operators' perspective. The author was the chairman of this vision group, which developed some scenarios for 3G for discussion during the GSM MoU Plenary in Hong Kong, autumn 1996. The discussions were too extensive to report in full here, but it is useful to report some of the conclusions looking to 2010.

The operators' vision is expressed through Figure 2.39. This speculates that telecommunications markets within the next 15 years will move from a very organised, regulated structure to a world with a few global players, probably organised in global alliances, and recognised by consumers as a set of global brands.

At the time of writing, these changes already seem to be happening. In most countries now a deregulated market has been established. Operators have built up alliances (e.g. FreeMove, an alliance of Orange, Telefonica, TIM, T-Mobile and others), with Europe leading in this field of globalisation and branding, some of which are regional and some have a global dimension. Furthermore, these business-level alliances are appearing to the users as brand wars, which have already developed in a major way in the past 2 years. The main difference between the vision (left-hand side of Figure 2.39) and today (right-hand side of Figure 2.39) is that the 'triad' of economic regions, Europe–Japan–USA, is much closer together than expected almost 10 years ago.

2.4.1.1 The Total Mobile Market by the End of the Decade

In contrast to the 3G top five countries described in Section 2.3.6.5, the total mobile market by the end of the decade 2010 will look a little bit different.

If the latest forecast numbers of global wireless subscribers are to be believed, (EMC, 2004b), the business expectations for mobile operators may be even more rosy than predicted. According to researcher EMC, the number of global wireless subscribers reached 1.5 billion in early June 2004. New numbers indicate that the global wireless market could attract as many as 240 million new subscribers in both 2004 and 2005, raising the total number of global mobile subscribers to 2 billion by as early as mid-2006. Much of this growth will come from emerging markets.

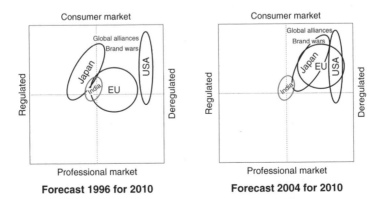

Figure 2.39 How the telecommunication market in 1996 was supposed to change for 2010 (left) and how it is seen today.

According to EMC, China will remain the world's leading wireless market, claiming 550 million mobile subscribers by 2009. The US wireless market is expected to remain the world's second largest market, growing from 157.3 million subscribers in 2004 to 223.9 million in 2009.

India is expected to quadruple its subscriber base to 117 million users in the next 5 years, raising it from the 13th largest wireless market in 2003 to the 3rd largest market in 2009.

Brazil is set to grow from 50 million wireless users in 2004 to over 100 million subscribers in 2008, making it the fourth largest global wireless market in 2009.

It is suggested that all of these figures include mobile and WLAN subscribers (or Wireless Fidelity (Wi-Fi), the American expression for WLAN, subscribers), because most mobile operators have already targeted the WLAN community and they add these subscribers to their total subscriber base.

The reason for the difference in numbers between the total market expectations and the 3G top five countries' expectations can be explained by the fact that in the developing and threshold countries, the 2G market has just started to explode. That is about a decade later than in Europe and Japan. These developing markets have a lot to catch up to meet the penetration rates of the established 2G markets. So it is no real surprise that China, India and Brazil will go to the top of the total mobile subscribers list.

2.4.1.2 WLAN and its Market Position versus UMTS/3G

It is difficult to say how the subscriber numbers of UMTS/3G and WLAN may split between the two groups. In this section a clue is given about the WLAN market opportunities as seen by the UMTS Forum in 2002. As usual, the UMTS Forum always takes a conservative approach.

In the days when UMTS was under development, people tended to talk about 'high-speed mobile data access' when they thought of business applications involving the transfer of large files, pictures, videos etc. These concepts were largely focused on business users and laptops. As the mobile market developed, there was an extremely rapid shift from business users to consumers, from practical applications to 'infotainment', and from laptops to mobiles and other handheld devices (PDAs). For business users and some consumers, these higher bandwidth capabilities are perceived as a solution to a real-world problem, while for other consumers they are seen as something that will enhance their quality of life. What was (and still is) needed are new ideas for 3G products, so-called 'killer applications'. Some possibilities could already be seen via the wireless access protocol (WAP), but that service was over-promoted and ultimately seen as having failed abysmally. But the truth is that WAP did not fail because the services were badly conceived – indeed, many of them were exactly what consumers wanted – but was branded a failure because the screens were too small, the user interface too fiddly and the transmission speed too slow. With the introduction of GPRS and CDMA2000 it worked much better, but people were still cautious.

For some technical enthusiasts WLAN or Wi-Fi came out of the blue when the industry developed wireless broadband data access at high speed for local islands (homes, buildings, offices, campus). This was a low power ($\leq 40\,mW$) short range ($\leq 150\,ft$) technology developed under the US Institute of Electrical and Electronics Engineers (IEEE) and named IEEE 802.11x-standard (see Section 4.7). First, it was created as a form of wireless access for computers/laptops used in bigger offices or on campus to avoid pulling so many cables in buildings and justify specific access handling. Quickly and cost effectively implemented on PCMCIA cards and more and more voluntarily installed on new CPUs (e.g. Intel's Centrino), it quickly exploited a gap in the market caused mainly by the absence of third-generation semi-mobile (nomadic) services.

WLAN has a lot going for it in the early adopter market. Most people having bought a brand new computer or laptop during the past two years are already WLAN equipped, and many have wireless routers at home. PDAs, too, are starting to come equipped with WLAN cards.

But while WLAN access as a technology is now proven, questions remain as to the geographic coverage and commercial access conditions. In other words, you have to figure out where to buy it, and who from. Looking at the UK market your access depends e.g. on whether you prefer your latte from Starbucks or your cappuccino from Costa Coffee. In the first case, you need to buy your WLAN hotspot access from T-Mobile; in the latter you need to buy it from BTopenworld. And nor can you wander into either chain confident in the knowledge that you will be able to browse the Web while you relax: access is available in selected branches only, and sometimes not even the staff can tell you whether you have struck lucky. Some operators and ISPs, e.g. BT, The Cloud and T-Mobile (Silicon.com, 2004), are now starting to offer customers WLAN roaming opportunities, which will exacerbate the above mentioned situation.

As mentioned earlier, WLAN is currently a very short-range technology. That means, when you are travelling, sometimes you still have to find the hotspot itself. It is no use turning up at an airport or railway station with your special subscription and your WLAN-enabled laptop, expecting to simply sit down and access your email. You have to find where exactly in the building the hotspot is to be found. Signposting normally does not appear to be a strong point for any of the public access networks, but that may change as more hotspots (e.g. business areas, industrial fairs etc.) become WLAN-equipped.

Summarising, it is not that easy to use WLAN as you normally use your mobile network. It does not turn up as 'always connected, anywhere, any time'. Another open problem is the security on WLANs. There are some security algorithms (wired equivalent privacy (WEP) or Wi-Fi protected access (WPA)), but they are insufficient for business use. A new security standard, the advanced encryption standard (AES), will be released this year, called WPA2, but it remains to be seen whether it has the same level of confidence as the GSM A5 or the UMTS equivalent security. Another open field is authorization, authentication, accounting (AAA), an item which is now properly solved in the mobile cellular world. From there, help may come: as more mobile operators recognise the possible value of WLAN access combined with their mobile net, the more existing AAA procedures will enter the WLAN world.

So, which technology will win? What does the interested industry need to do to succeed? The truth is that neither 3G nor WLAN hotspots have yet truly become mass-market services. They will not succeed until they can break through the mass of confusion and apathy that exists in both consumer and B2B markets. Part of the solution to that is to improve dramatically the user-friendliness of things like wireless routers. Gaining mobile access to the net must be as simple as buying a card in a coffee shop and logging on while sipping a cappuccino. Another key part of the solution is not to compete with the other technology, but to offer seamless access.

A very important step forward is the introduction of combined cards offering GSM/GPRS, UMTS and WLAN access: 'We'll give you mobile net access, and let us worry about the method'. That has started in April 2004 and is an important step forward to meeting customers' expectations. It does not matter whether you are at home, in a coffee shop or parked in a layby on the motorway, you get your access to your email and the Web 'any time, anywhere', as promised in the early days when the UMTS Task Force offered their vision of the mobile broadband world to the European Commission in 1996. The transition between fixed and mobile, and between Wi-Fi and Bluetooth-enabled handset, should be transparent.

Consider what Ben Lovejoy (2003) mentioned in his article '3G and Wi-Fi: friend or foe?':

'The fact that you've invested millions in licences and infrastructure, and have investors demanding to know when the money will start flooding in, doesn't mean that we can somehow magically skip straight to the mass-market phase. Early adopters are the people we need to woo first, and what they tell us they need is access to the net, anytime, anywhere. That's where the initial money will come from, and – more importantly – what will ultimately lead to mass-market adoption of both technologies.'

Last but not least it should be mentioned that ETSI has also developed a WLAN standard, named *HiperLAN*. In contrast to its American rival, it is more complicated and complete as it offers security, handover and roaming procedures from the very beginning, but on a proprietary basis. It is hard to say whether it will enter the market successfully now that IEEE 802.11x has already become a de facto world standard for WLAN.

Market Expectations and Operator Revenues for UMTS and its Main Competitor WLAN

With respect to the success of WLAN in the markets, many organisations, consultant companies, operators and vendors have tried to analyse the market potential of WLAN in the new UMTS/3G environment. It was very logical that the UMTS Forum, as the cross-industry organisation for 3G, researched the market potential on behalf of its members. As follows, the results of this study will be reported, compared with other research and put into context regarding the economic environment in 2004.

The Forum study built on its 3G market research on the potential revenues for operators. Although the Forum usually looks at the long-term revenues (10 years), this time it took a 5-year forecast due to an urgent request from its members. First, it analysed the potential customer base for WLAN compared to that of UMTS/3G. It emerged clearly that the main target group for mobile data communication usage would be the 'mobile workers', those people travelling very often and needing mobile data access any time and anywhere. These people are called 'nomads' and they differ from those who travel frequently but need mobile access only at dedicated places (hotspots). Using the data of the ILO, a subsidiary of the UN, it was found that out of 1.8 billion workers worldwide, 680 million could be expected to be at least occasionally mobile workers. Of those, 104 million mobile workers, including tele-workers, business travellers and remote workers, could be seen as potential users. A fifth (20 million) of those could be seen as WLAN users (Figure 2.40).

The next step should be to find out how many of these public wireless users would be WLAN users only and how many would use WLAN or 3G, depending on the availability of a network. The result was that about a quarter, or 5.3 million people, would use both networks depending on availability. The remaining 15.2 million would be WLAN users only, including occasionally users and business executives who will use WLANs less frequently. Pure 3G users are people who need spontaneous access to mobile networks any time, anywhere to get information immediately which is important for their present work – that is, people

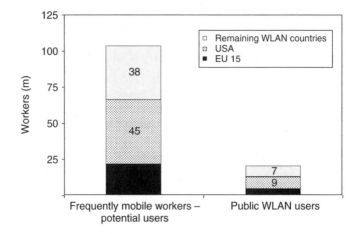

Figure 2.40   Public WLAN addressable market 2005. *Source*: Telecompetition. © UMTS Forum, 2002.

working on areas like field sales, technical and engineering services, who more often work out of their homes or vehicles or conduct their work activities in non-public, local areas. Those customers are less able to use hotspots frequently and would use WLAN only exceptionally; they just login either when they know they have access in that place or if it will work automatically without worrying about which method is best at any given location.

The UMTS Forum's market studies on 3G revenues for operators in 2005 (UMTS Forum, 2002c) forecast worldwide operator-retained revenue for mobile intranet/extranet access to be $9.8 billion. As shown in Figure 2.41, the 20.5 million mobile business workers using

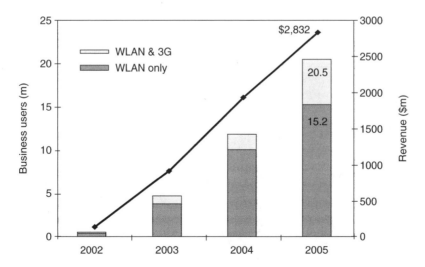

Figure 2.41   Public WLAN business users and revenue forecasts 2005. *Source*: Telecompetition and UMTS Forum, 2002.

public WLAN, of whom 5.3 million will use WLAN and 3G, may generate about $2.8 billion in WLAN service revenues by 2005, which will represent a total of about $12 billion for the combined market.

The 5.3 million nomads would generate less than 3% of the total forecast 3G revenues in 2005 of $67.7 billion (UMTS Forum, 2001c, table 1; see Figure 2.29). These WLAN users may be seen as early adopters and they will cover a quite interesting niche market which operators cannot ignore.

It is interesting to compare these numbers with a study by Analysys Research in 2002 and 2003 (Figure 2.42).

The figures published in 2002 by Analysys and the UMTS Forum are very close. Regarding the remaining difficult economic situation in 2003, almost the same figures are seen in the Analysys report, but the curve moved 2 years further down the road. The reader should not be surprised if another shift of a year or so follows, comparing the data with those received by some other companies and their data: Probe group (RCR Wireless, 2004) forecasts the number of mobile workers using WLAN in 2008 worldwide to be 600 million (compared with the UMTS Forum's 680 million in 2005). With respect to a press release at CeBIT 2004 (FierceWireless, 2004a), it is said that T-Mobile generated about $1.4 million in revenue per month in 2003 from its US network of 4200 hotspots. That comes to about $13 per day per hotspot, or $400 per month per retail location. As Probe says, 'This is hardly enough money to even pay for the service costs, much less to generate a profit'. It seems that there is still a long way to go until WLAN as a stand-alone service becomes a profitable business. But, as was said earlier, mobile operators are going to merge both businesses and make it look like a single wireless business to the customer.

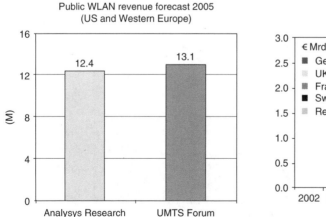
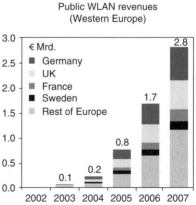

Figure 2.42   Public WLAN business users in Western Europe and the USA, Forecasts 2005. *Source*: Analysys Research and Telecompetition and UMTS Forum, 2002 (left) and Analysys Research, March 2003 (right).

2.4.1.3 Financial Aspects

A fundamental assumption underlying this study is that 3G will essentially free mobile networks from capacity constraints. The main factor behind this assumption is the inherent efficiency of all-IP packet-based networks. The identification of additional spectrum at WRC-2000 supports this assumption. However, significant amounts of additional spectrum may not become available in some regions until mobile service providers can demonstrate the need for it to be released for 3G. And in the USA, service providers will have access to very little additional spectrum compared with other regions.

Capacity is not currently a significant issue in the USA. In such circumstances the competitive nature of the mobile environment creates downward pressure on pricing and a decline in ARPU. Almost all 3G service providers will launch services in an ultra-competitive environment that will be focused more on data than on voice services. Mobile data is universally regarded as the mechanism for reversing the downward trend in ARPU.

Many 3G service providers will also be burdened with up-front investments in licence fees much higher than anticipated in their original business plans. The need to recover this investment as swiftly as possible has shifted the emphasis of many service providers from the business market to the consumer mass market for early 3G services. The competitiveness of the environment increases substantially with such a shift.

First mover advantage will be a paramount consideration. This is not a mere reflection of the Internet 'land-grabbing' approach that sees the acquisition of market share at any cost as the means to build long-term advantage and (ultimately) revenues. First mover advantage will be important in the 3G mobile data environment because of the personalised nature of the key 3G services, which are designed to generate stickiness and minimise churn. That will increase the competitive nature of the environment even further.

This forecast shows significant revenue potential even though it assumes average mass-market pricing. Over the next 5 years, an element of mass-market pricing could be introduced to replace premium pricing in some regions. The average prices assumed in this study are compatible with a range of service provider pricing strategies.

2.4.1.4 Trends in Mobile Data Traffic Pricing

The recent Ovum Research study (2004) on this subject discusses the question 'Trends in mobile data traffic pricing: who to charge? What? When?'. In summary it says:

There is an inherent conflict between the buyers and sellers of mobile data services. Consumers want simple pricing plans and no hidden extras, while operators need to be able to charge for the use of their finite amount of spectrum. Whilst consumers don't know or care how much data it takes to download a weather report, or the latest football scores, mobile operators know that 'not all data transmissions are created equal.' Network operators must balance these two competing requirements and square the circle.

The consumer market focuses on three main types of data:

- messaging;
- browsing and server side content access;
- content downloads.

Each of these requires different approaches to pricing, and what works in one part of the world may not always work in another. An example of this can be seen with recent messaging trends. When photo messaging was introduced in Japan by J-Phone (now Vodafone KK), there was a tariff based on the volume of data carried. In 2002 when other operators around the world were offering this service it became apparent that consumers in their markets were not comfortable with this, and wanted event-based pricing:

- a fixed uniform price per message;
- no additional charge for the GPRS traffic used to send and deliver the message;
- sender pays, not receiver.

2.4.1.5 Non-Voice Services

Up to 2000 data services for the mass market were virgin territory for the mobile industry. The industry had failed to make a success of mobile data services targeted at the business community. But the industry had been pleasantly surprised by the unanticipated success of consumer mobile data services such as SMS and i-mode. Building on these successes will be the key to 3G.

One problem affecting the construction of successful mobile data services is user expectations. User expectations are still being set by the fixed-Internet experience in those countries with high Internet penetration – and that is a rapidly moving target. Another problem is that the addition of mobility has become part of user expectations for 3G compared to the experience of the fixed Internet. But paying a higher premium for this addition of mobility does not figure prominently among user expectations. Compromising on functionality in the mobile environment is not a market expectation. Mobile service providers will be expected to offer mobile ISP-type services, but it is hard to envisage profitable revenue being generated through this route.

Profitable revenue should, however, come from services that really take advantage of the unique characteristics of mobile and that have no immediate counterpart in the fixed environment. Customised infotainment services through mobile portals, multimedia messaging services and location-based services seem to be the key weapons in the mobile service provider's armoury. But video phones may drop into this phalanx with greater success (see Section 2.4.2.2).

Relationships with partners will be key to success, particularly in the all-important mobile portal arena. Earlier predictions that content and media owners would grab and dominate this space are not true. This should not be a surprise. Mobile is only one channel to market for content and media owners and is therefore not a channel to be acquired 'at any price'.

Partnerships between content/media owners and mobile service providers that capitalise on each party's strengths are the key to success, whether such partnerships are organised through joint ownership ventures or through mechanisms such as mobile virtual network service providers.

2.4.1.6 Roaming

Another dimension vital for success is the ability to roam with full data service capability in an international or inter-regional environment. Without such capability, mobile data will be constricted to a cordless or wireless access service. The need for roaming in the data environment places great demands on service portability and raises yet unresolved issues concerning billing and tariffing across service providers.

It is clear that very significant advantages can be gained by those service providers who can construct a pan-regional presence (either through ownership or partnerships) and can integrate business procedures and commercial offerings to deliver seamless service to subscribers across national or regional boundaries.

The lack of a single global standard for 3G will complicate the situation for manufacturers and users. Paradoxically, however, it could benefit UMTS. Only a limited number of the multiplicity of multi-mode and multi-band terminal combinations that result from the family of systems solution in IMT-2000 and the spectrum bands identified at WRC-2000 will find their way to market with sufficient economies of scale to permit acceptable pricing. This will automatically support the already dominant players – the strong will get stronger and the weak will get weaker.

Service portability is of course also a factor in roaming between 2G and 3G networks. The regulatory intent to support new entrants in the early stages of network rollout is already evidenced by the requirements for national roaming imposed in certain countries. Wide coverage for data services is a particularly important requirement in the marketplace that will be intolerant of any significant service degradation suffered when moving out of 3G coverage. The consumer market is notoriously unforgiving and has a track record of condemning services whose availability is limited at launch. It is hard to recover from negative perceptions.

2.4.1.7 Importance of Terminal Novelty

Taking into account the Ovum Research (2004) study, operators need to analyse very carefully customer needs and expectations and react very quickly to their demands. Otherwise a very successful model, once introduced in a new market, but not adopted in time with customers' expectations, could fail quickly, as happened recently (first half of 2004) with Vodafone KK (*Wall Street Journal*, 2004), when they spent too much time trying to build a global brand and too little time giving their subscribers what they wanted. After Vodafone's very successful introduction of *Sha-mail*, the mobile picture phone service in 2002, analysts and insiders today claim that they missed introducing lots of 'cool' new phones with new innovative mobile services such as videophones and 3G afterwards. Old phones are seen as boring and out of date, something catastrophic in a wireless market dominated by novelty and new technology.

2.4.2 CHANCES FOR 3G

2.4.2.1 Service Demands

The presence of an air interface together with the limited availability of spectrum creates inherent bandwidth/capacity constraints in current mobile cellular systems. Speed limitations in mobile cellular will continue to limit parity with fixed access speeds.

The market study acknowledges all these issues and has still forecast impressive growth in revenues and subscribers for 3G networks. The challenge now lies with the service provider to deliver services in this new IP-based world. Table 2.8 summarises the analysis of some of the technical characteristics of the services studied.

Table 2.8 Some distinguishing technical characteristics by services

Characteristic	Mobile Internet access	Mobile intranet/extranet access	Customised infotainment	Multimedia messaging service	Location-based services	Rich voice
Connection type	1:1 1:many (email)	1:1 1:many (email)	Primarily 1:1 Broadcast can be 1:many	1:1 1:many	1:1	1:1 1:many
Portable versus mobile	Portable	Portable	Mobile	Mobile	Mobile	Mobile
Latency	Latency insensitive	Latency insensitive	Primarily latency insensitive	Latency insensitive	Latency insensitive	Real time
Person to person				Yes	Yes	Yes
Person to machine	Yes	Yes	Yes		Yes	Limited
Machine to machine						
Addressing and call control	IP and domain names	IP and domain names	IP and domain names	IP and domain names, session initiation protocol (SIP)	IP and domain names	E.164 telephone numbers, SIP/H.323[a]
Access	Always-on	Always-on	Always-on	Always-on	Always-on	By invitation
Speed requirements	Medium to high	Medium to high	Low to medium	Low to medium	Low	Low, but high for video
Service symmetry	Asymmetric	Asymmetric	Asymmetric	Asymmetric	Asymmetric	Symmetric

[a] E.164 and H.323 are ITU-T Recommendations (see Chapter 4).

Source: Telecompetition Inc., September 2000 with additional analysis by author.

Table 2.8 shows how different rich voice is from the other services – service symmetry and latency being the key distinctions. Both these characteristics lead to the circuit-switched, permanent connections with call set-up that are the hallmark of telephony networks. A speciality is the videophone service, which will be explained later in detail (see Section 2.4.2.2).

The other services, both those that are Internet-centric and those that take advantage of the always-on characteristic of IP, are much more similar.

In June 2004 Teleconomy published a survey of 1400 people in the UK, suggesting that teenagers, not business people, could be the lifeline for mobile operators looking to make revenues from the data services market, and that children between 10 and 14 years would be the mobile phone buyers of the future and the biggest users of mobile data. Three in four of 'M-Agers' (that is, the group of 10–14-year-old) are aware of mobile calling applications, compared to one in two adults. Two thirds of them are savvy about the benefits of Java applications, compared to less than half of adults. It is not said how much ARPU they may generate compared to adults.

2.4.2.2 Video Services

A viable business model for video distribution and content protection needs to evolve. Interoperability, interconnection and roaming issues are to be fully defined, and standards across the value chain are required to enable operators to build a full service that includes content, servers, applications and handsets.

Despite these initial barriers, an end-user survey in the USA of In-Stat/MDR and ARC Group (IEE Commentary, 2004) suggests that 13.2% of US wireless subscribers are extremely or very interested in purchasing video services for their wireless phones. The analyst suggests that this represents a natural demand for mobile video services, prior to any large-scale carrier deployments or market messaging.

This interest in mobile video is higher than for all other prospective mobile multimedia services, such as gaming and music services. In-Stat expects that in the USA alone, 22.3 million people will be viewers of mobile video content, and 31.1 million will use video messaging services, and that mobile video services will account for approximately 14.9% of total wireless data revenues.

According to the latest report of ARC Group, the forecast mobile video market will generate worldwide revenues of US$5.4 billion in 2008. ARC forecasts a quite slow rate of adoption between 2003 and 2005, but from 2005 onwards strong growth is anticipated until 2008. The Group expects for 2008 about 250 million subscribers worldwide for mobile video services, assumed to be the biggest application category in 3G services (www.fiercewireless.com, www.3g.co.uk). That seems to be quite likely, if you take into consideration the latest research of the UK Teleconomy Group.

Video download is expected to be second to video messaging in terms of users until 2005, when video streaming will take over second spot, based on the higher penetration of 3G networks. Streaming is expected to be a preferred method of consuming video content, since it has a much more immediate viewing experience than video download, and enables longer video clips and also TV-like live broadcast services.

2.4.3 BARRIERS FOR 3G

2.4.3.1 The Late Start of UMTS/3G

At this point, it is time to make a short comment regarding the late start of UMTS/3G. Some of the critical success factors for the start of UMTS/3G are:

- a successfully interoperating 2G/3G network;
- the availability of customer demanded services and applications;
- the availability of a sufficient number of handsets/terminals.

Unfortunately, neither operators nor vendors could keep the predicted starting time of early 2002 in Europe, although NTT DoCoMo had already successfully started its *Freedom of Mobile multimedia Access* (FOMA) system in late 2001. To be honest, it was not as successful as expected. The reasons were multi-fold. First, it was said that the terminals were too bulky and the battery life too short. Second, the coverage for FOMA was poor compared to their second-generation system i-mode. And third – the most critical factor – almost all services of interest for consumers were available on i-mode, except picture services and video, and that was not enough to justify a change to FOMA. Lastly, its main competitor, KDD's multimedia service, based on CDMA2000 technology, was cheaper and more successfully introduced to the Japanese market, but eventually gave FOMA a second wind.

In 2001/2 many American and European journalists were moaning for obvious reasons about the quality of the FOMA system and compared its hampered start with the less successful start of WAP in Europe. Some were keen to support CDMA2000, others feared a similar disaster as seen with WAP in Europe. With reference to these experiences the big operators delayed the start of UMTS/3G again and again, because the critical success factors mentioned above had not been solved satisfactorily. Only in the first quarter of 2004 did they decide to offer 3G services to their customers, starting with pure data services, and followed with full mobile services (including voice) in second quarter 2004. However, some courageous operators, like Hutchinson 3G (UK and Italy) and Mobilkom (Austria), did an earlier start, but suffered from a shortage of terminals.

2.4.3.2 Camera Phones Ban

Soon after the introduction of camera phones people misused them to take pictures in private or secret areas. That led to a ban on using camera phones inside companies to protect research and development, but also strategy and financial plans. On the other side in some Islamic countries the use of mobile phones was banned or at least restricted for religious reasons. The problems that camera phones can generate are described in more detail in Chapter 7.

Another example of a ban on camera phones is actually on its way in the USA, where the Administrative Office of the United States Courts is preparing a decision on whether camera phones should be banned in courts (CNet News and CTIA Daily News, 2004a). At present, recording devices in general are not allowed in courtrooms to protect witnesses and minors. Camera phones are a concern because the devices are more easy to conceal. Some courts already ban the phones and require that phones be turned in or shut off in the courtroom.

2.4.3.3 Health Issues

One of the most recognised publications on this subject is the Stewart Report commissioned by the UK government in September 1999 (IEGMP, 2000). As the use of mobile phones and related technologies will continue to increase for the foreseeable future, its recommendations are:

- The balance of evidence to date does not suggest that emissions from mobile phones and base stations put the health of the UK population at risk.
- There is now some preliminary scientific evidence that exposures to radiofrequency (RF) radiation may cause subtle effects on biological functions, including those of the brain. This does not necessarily mean that health is affected but it is not possible to say that exposure to RF radiation, even at levels below national guidelines, is totally without potential adverse health effects.
- The Expert Group has recommended that a precautionary approach to the use of mobile phone technologies be adopted until more detailed and scientifically robust information becomes available.
- For base station emissions, exposures of the general population will be to the whole body but normally at levels of intensity many times less than those from handsets.
- Some people's well-being may be adversely affected by the environmental impact of mobile phone base stations sited next to houses, schools or other buildings, as well as by fear of perceived direct effects.
- For all base stations, including those with masts under 15 m, permitted development rights should be revoked and the siting of all new base stations should be subject to the normal planning process.
- The use of mobile phones while driving can have a detrimental effect on the quality of driving. Drivers should be discouraged from using mobile phones while on the move.
- The widespread use of mobile phones by children for non-essential calls should be discouraged.

This report was a milestone for the whole mobile industry. Its recommendations have been taken very seriously and parts of them have been incorporated into legislation, e.g. the ban on using mobile phones while driving without hands-free kits.

In May 2004, the *Neue Zürcher Zeitung* (NZZ) reported a tremendous threat for UMTS networks in Switzerland caused by an MP of the Swiss Green Party and an MP from the Swiss Conservative Party, who are postulating to reproduce the Dutch Study of TNO (Zwambarn *et al.*, 2003) for Switzerland and to introduce tougher hazard emission limits as used in other countries throughout Europe. However, the TNO study must be seen in the light of the ongoing total worldwide research into possible health effects of mobile phones and base stations. The authors of the Dutch study themselves acknowledge that this work was based on a small sample size and will need to be repeated by an independent research institute. The Dutch Government explained in its statement that at this point the results of the study cannot lead to final policy conclusions. The NZZ report does not take into account that the Federal Office for the Environment (BUWAL) has not asked the regulatory authority (Kommunikationskommission) to change current Swiss limits. If the Swiss replication study on 3G and well-being confirms the Dutch TNO results, a change of Swiss UMTS licence conditions, lowering the current

Swiss limits on radiation, would jeopardise operators' investment tremendously. Presently, there is no moratorium initiative undertaken by the Swiss parliament. There seems to be a mutual understanding in Swiss politics to tighten UMTS limits if the Swiss replication study confirms the TNO's findings in late 2005.

2.4.3.4 3G Business Development in a Worldwide Context

During the 1990s, Europe led the world in mobile telecommunications, as its successful adoption of the GSM standard led to the widespread take-up of mobile telephones. But now the situation seems to be changing: Europe is falling behind, Japan has taken the lead and North America is speeding up in deploying national high-speed cellular networks, Wi-Fi hotspots and devices such as the Blackberry and wireless PDAs. Europe is only now starting to catch up, with the recent introduction of UMTS services for laptops and Wi-Fi deployments in key economic areas by the leading operators, like Vodafone, T-Mobile and TIM. Initial deployments of such services have primarily been in North America and the Asia–Pacific region rather than Europe.

That leads the European Commission to more political activity to pave the ground for better telecommunication business opportunities in the European Union. But there is still a long way to go. In this sense, Europe has to speed up, if the European-born UMTS is to slip successfully into the shoes of the worldwide prosperous GSM.

3

Services and Applications

3.1 INTRODUCTION

3.1.1 THE 3G BUSINESS CHAIN

Technological and commercial developments are melting together information, communications, commerce and entertainment into one large, consolidated industry. Part of the reason for this evolution is that more consumers are accessing the Internet using multiple devices and multiple communications networks. Consumers are changing their behaviour and consumption patterns, and mobile communication has become a daily necessity. In addition, new tools and facilities are becoming available that improve the consumer Internet access experience. These factors have created a number of opportunities and challenges for new and existing businesses to develop their Web presence to serve their customers better.

Wireless access to the Internet will drive the overall development of the Internet for several reasons:

- Access service providers and Internet businesses will increase their mobile presence by offering e.g. traffic navigation services and location-based services such as restaurants, hotels etc. The increased consumption of such services will drive their future growth.
- The mobility and immediacy offered by wireless access allows Internet content delivery and commerce to be location-specific as well as non-location-specific.
- The person-specific nature of wireless access allows companies to develop customer profiles that enable them to narrowcast and target more accurately the distribution of value-added information to customers.
- Location-based facilities and services provide another tier of customer knowledge that allows Internet businesses to deliver 'context'-specific services that also improve customer value.

The Mobile Multimedia Business: Requirements and Solutions   Bernd Eylert
© 2005 John Wiley & Sons, Ltd

In short, wireless access to the Internet is an opportunity for all Internet-related businesses to learn more about their customers, understand their consumption patterns, strengthen their customer relationships and provide more personalised services. These are critical components of Internet business strategies on which wireless operators and service providers will base their full Internet solutions.

Services such as traffic location and navigation services and location-based services affect the 3G business chain in terms of both technical standards and commercial aspects. Figure 3.1 illustrates these two parallel axes of the business chain. This subject will be discussed in more detail in Section 4.3.

3.1.2 THE MOVE TO MULTI-SERVICE MARKET OPPORTUNITIES

The new business opportunities facilitated by UMTS will add new market segments to existing and traditional telecommunications markets. UMTS will offer:

- fast mobile multimedia capabilities;
- service portability;
- personalised and ubiquitous communication capabilities.

The question arises as to which services will become most popular, reflected by the most asked question from journalists and analysts: 'What will be the "killer" application?' So far, it appears that there is no single one, but rather some candidates which may become the leading applications. As mentioned in Chapter 2, the videophone is a strong contender.

To better understand the requirements and dependencies of the services categories in Chapter 2, some representative examples reflecting the benefits of UMTS for end-users have been chosen and will be described in detail here.

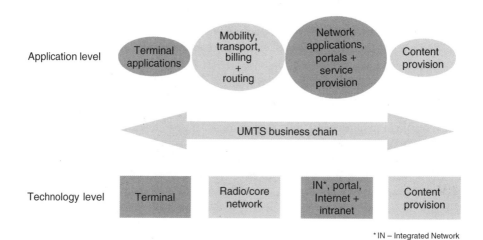

Figure 3.1   Correspondence between applications and technology in the 3G/UMTS business chain.

3.1.2.1 Virtual Home Environment

It is highly desirable that mobile workers are able to obtain their personal service profile at any place in the world independent of the actual network operator. The proposed concept of virtual home environment (VHE) in 3G systems leads to the portability of services over 'borders' (networks, countries, operator domains). This enables mobile workers to achieve the same productivity anywhere, whether it is at home or when travelling, thus losing no productive time as a consequence of having to undergo 'training' periods at every new location. VHE will therefore become a means of increasing staff efficiency.

The right 'service bundle' from narrowband (voice only) to broadband (multimedia) services combined with the integration of global mobility in existing processes such as workflow and e-/m-commerce will give new benefits in well-known processes or make new processes possible for the first time.

A lot of time and money will be saved with the introduction of the VHE concept if e.g. employees are able to use (all) services at any place in the world in the same manner that they are accustomed to at home (service portability). All of these opportunities lead to advantages in the competitive environment to provide additional business opportunities for both operators and enterprises/organisations.

But this service is not only an advantage for business people; it also has a lot of benefits for ordinary consumers who travel infrequently, as they can use all the same services with the same look and feel in a visited network as the one they use at home.

3.1.2.2 Key Market Segments

Today, there is already a blurring of traditional business/consumer market segmentation and this will be even more the case tomorrow. It seems to be easier to think in terms of lifestyle management rather than business/consumer market segmentation. This subject is considered in more detail in Chapter 7 from a sociological point of view.

The most important lifestyles are as follows:

- Business professional, product managers, self-employed (e.g. small office/home office): high-value mobile users (e.g. busy decision makers), users with specific occupational requirements for high volumes of information while mobile (requiring remote and mobile access to corporate and external information). Typical services required: intranet access, messaging and scheduling systems.
- Young generation: often early adopters of technology. Typical services required: messaging, games, entertainment-oriented services, video phoning.
- Family:
  - Parents: especially, if both parents work they would share the responsibilities of maintaining the home and of childcare and keep in touch with each other. Typical services required: messaging monitoring.
  - Children: ability to communicate with their parents. Typical services requried: e.g. messaging.

- Senior citizens: more reliable support electronically and reduction of requirements for labour intensive support services. Typical services required: e.g. medical monitoring, location-based medical and social service.

3.1.2.3 Trends with 3G Mobile Services

Many services, although they may start with 2G, will become more affordable using 3G. Capabilities such as video-on-demand or the transmission of large data files containing detailed graphics and descriptions will need broadband. While technically possible over narrowband networks, they would be very slow and not acceptable to any business organization.

Services which already exist will be greatly improved by the provisions of interactivity and mobile multimedia based on customer segmentation according to lifestyle management. Traditional business/consumer market segmentation may blur, because it is the same person communicating, whether synchronising his/her schedule or sending a message from home. Some analysts see work-oriented applications driving the 3G market in Europe and in the USA, whereas in Japan the high value placed on increasingly more sophisticated consumer-oriented equipment could have a substantial impact. The demand for increased personal productivity is also of importance. The blurring of boundaries between work and home will continue.

3.1.3 QUALITY OF SERVICE

In Section 3.1.2 and earlier, the term 'services' refers to applications as defined in Section 2.2.2.1. However, the term 'service' and particularly the concept of 'Quality of Service' (QoS) has a very different meaning, which is relevant to UMTS/3G networks.

QoS is a technical term that specifies data transmission throughput in a network. Various QoS levels may be offered to the customer; for example, best effort, differentiated or guaranteed.

In the remainder of this chapter, the relationship between a given QoS and the application being delivered will become more obvious.

3GPP[1] has defined four preliminary QoS classes for data transport over UMTS (3G) systems:

- conversational class
- streaming class
- interactive class
- background class.

The conversational class in general sets the highest requirements for QoS, whereas the background class requires the lowest and can be seen as a best effort type. As a general requirement it can be seen that QoS parameters should not be restricted to one or only a few mechanisms, but provide capability of different levels of QoS by using UMTS (3G) specific control mechanisms. This is especially true for the conversational class, but also in regard to inter-working with other types of networks. In addition, QoS for UMTS (3G) has to consider the efficient use of the radio spectrum. It has to allow an independent evolution of the core networks and the radio access networks. The evolution of UMTS (3G) QoS should be independent as much

---

[1] 3GPP (Third Generation Partnership Project) is an informal cooperation of six regional SDOs of Europe, Asia and North America to standardise ITU-R IMT-2000 access technologies IMT-DS, TC and SC; its counterpart for ITU-R IMT-MC is called 3GPP2 (for more see Chapter 4).

Table 3.1 Quality of Service (QoS)

Type	Service	Data rate (kbps)	Delay	Delay variation	Reliability
Conversational/ real time	Conversational voice	4–25	<150 ms	<1 ms	<3% failure error rate (FER)
	Videophone	32–384	<150 ms		<1% FER
	Telemetry (control)	<28.8	<250 ms		~0% FER
	Games	<1	<250 ms		<3% FER
Interactive	Voice messaging	4–13	<1 s	<1 ms	<3% FER
	Web browsing		4 s/page		
	E-commerce		4 s		<0% FER
Streaming	Streaming audio	32–384	<10 s	<1 ms	<1%
	Video	32–384	<10 s		<1% FER
	Telemetry (monitoring)	<28.8	<10 s		~0% FER

© UMTS Forum (2000b).

as possible from the wireline QoS development, but has to consider the requirements of inter-working with those wireline services allowed by UMTS (3G).

Table 3.1 shows a preliminary QoS scheme for UMTS (3G) services, which includes inter-operability with second-generation high bit rate packet-based solutions. The full utilisation of the UMTS (3G) bandwidth, especially for conversational (real-time) types of service, will require further development.

Enabling support for voice and other real-time services along with data capabilities requires QoS features to arbitrate access to the limited transmission resources in the wireless environment. The wireless and mobile environment requires QoS support for the last leg between the mobile device and network access point, accommodating roaming and unique characteristics of the wireless link. When selling differentiated services to the user, the actual 'product' (e.g. content) takes the form of a service level agreement. For the operator, the first step is to define the class of service, the standard set of features available within it.

QoS, on the other hand, is defined as a set of performance attributes associated with the service, such as availability, delay variation and the throughput of packets and packet loss.

3.2 DESCRIPTION OF SOME SERVICES AND APPLICATIONS

In general, basic service concepts are common between 2G and 3G. However, the service delivery mechanisms and user device attributes and interfaces will be vastly improved with 3G. The applications being developed for the Web and intranets will be a key source for 3G mobile applications. Many people will prefer to access the information superhighway via mobiles rather than personal computers. Most mobile users will use the Internet in different ways from personal computers users: they will go for short messages and quick transactions rather than leisurely browsing.

Mobile commerce should not be thought of as a service in itself, but rather as a generic set of functionalities enabling many services like m-brokerage, m-shopping, m-auctioning, m-banking, m-cash etc. Some typical services are described briefly.

3.2.1 ENTERTAINMENT

3.2.1.1 Location-Based Entertainment

Description
Location-based entertainment provides value for an end-user in certain locations. In this case value relates to excitement, pleasure, rewarding experience and more tangible things like information, speed and convenience. Location-based entertainment is part of location-based services. It has started on 2/2.5G networks already, but will improve in quality and ease of use on 3G.

Facilities
Location-based entertainment allows the customer in certain locations to:

- check programmes in theatres, cinemas etc.
- make ticket reservations (theatres, cinemas etc.)
- make travel reservations
- play local games or join a local game competition
- gamble on sports events, e.g. guess the team player who scores the first goal
- get local tourist information
- get information about local restaurants, discos etc.
- receive publicity on events in the area
- get information on sports in the area
- participate in local chat groups, e.g. backpackers arriving at a new place.

Importance
3G networks will offer excellent opportunities for location-based entertainment due to higher data speed and more advanced mobile terminals. Content will be richer and there will be access to the Internet in 3G. Some offers have already started on 2/2.5 G networks, e.g. the tourist guide in Rome, Italy, where users can receive location-dependent information on tourist attractions on their mobile phone.

Usage
Usage will depend on the location. Users will be more active in the places with a larger variety of different location-based entertainment services. Location-based entertainment services related to tourist attractions will be used frequently.

User Segmentation/Profile
The target market for location-based entertainment is the consumer segment (private individuals). The detailed segmentation is:

- all ages, from very young kids of 10 to older globetrotters of 60+
- people familiar with Internet services
- people familiar with mobile services.

Duration
The duration of each session will be around 2 minutes with a data volume specific to service.

Number of Sessions
Location-based entertainment services are easy to use and the consumer segment will use them on average five times a week. Some users might use the service several times per day.

Benefits
Three groups will benefit from this service: customers, service providers and operators. This will be an attractive offer in the market for numerous reasons.

- Customer benefits:

    - easy access to services in the area
    - time saving
    - up-to-date information
    - entertainment
    - profiling, i.e. user specifies his/her own preferences for the location-based services.

- Service provider benefits:

    - additional distribution channels (e.g. tickets)
    - advertising potential
    - attract new potential customers
    - possibility to offer services to different target groups.

- Operator benefits:

    - supports new business models
    - increased airtime usage
    - possibility to use zone tariffing
    - possibility to offer services to different target groups
    - potential to attract new customers.

Expectations
The expectations on this service will be numerous. Most of them will be visible to the end-user directly.

Drivers
- trend of strong growth of telephone and Internet services
- growth of mobile phone penetration
- customer requirement for easier and faster services
- customer requirement for services 'here and now'
- customer requirement for more value (excitement, pleasure, rewarding experience and more tangible value like information, speed and convenience).

Inhibitors
- people are not willing to spend more money for this service than for traditional services (location-based entertainment has to be cost-effective)
- service is not easy to use
- service is not trusted.

Source of Revenues
- advertising as an important source of revenue in location-based entertainment
- monthly fees to get location-based entertainment
- airtime used
- content-based charging.

Market Time Scales
Location-based entertainment is already available and will evolve with standards, technology and handset availability. Functionality and availability of this service affect maturity year by year.

3.2.1.2 Interactive Games

Description
Interactive games are an example of the type of mobile infotainment available on modern mobile handsets. They allow the player to download new games or levels, compete against other remote users and upload their highest scores, etc.

Facilities
Although interactive games are not broadly available in telecommunications today, 3G will enhance and develop their interactivity and multimedia content. The key changes to games that will be seen as new technology is implemented are:

- improvements to user interface (i.e. use of audio and video)
- interaction with other players (group gaming)
- ability to download new games and game upgrades over-the-air or via kiosks
- security improvements will allow gambling, lottery and competition type gaming to use the wireless platform.

Importance
In general, interactive gaming is not an essential part of life. However, examples in the Asian-Pacific markets show that especially young people do have a thirst for these new entertainment elements. Operators and service providers who are keen to follow their customers' needs will have to offer such services as interactive games as well.

Usage
User segmentation/profile: the games will be targeted primarily at the 6–35-year-old market. Gamers are becoming older and the women's market has already grown by over 30%. If some games are downloaded onto new mobiles as standard, the market growth will mirror the general upward trend in terminal sales.

Duration
Typical duration is 5–40 minutes per game, perhaps longer, depending on the level of interactivity.

Number of Sessions
This depends on the nature of the game and on how hooked players become!

Benefits

Three groups will benefit from this service: customers, games developers and operators. It will be an attractive offer in the market.

- Customer benefits:
    - opportunity to use leisure time for fun
    - simple access in any environment to familiar games from other media
    - creation of virtual gaming communities
    - win prizes
    - escape from the real world.

- Game developer benefits:
    - act as a channel to subscribers
    - increase market share (appeal to new customers) and customer penetration
    - increase revenue.

- Operator benefits:
    - customer loyalty and affinity
    - greater use of network; new revenue streams
    - make the phones more fun.

Expectations

Within a growing market interactive games will play a significant role. The impulse for this service will come from Asia-Pacific and inspire especially young people, who already see mobile devices as a communication device which includes all kind of entertainment, e.g. as game boxes as well (Digital World Research Centre, 2004).

Drivers

- growing market
- the ability to increase the standard of graphics/interactivity to be able to rival that of desk-based PC games.

Inhibitors

- download speed on a handheld device
- battery power/consumption
- noise or neighbour annoyance in public places
- operator connection charges
- reliability of the connection
- requirement and improvement of concepts and prototypes for devices.

Source of Revenues

- the initial game download
- percentage of data download/airtime revenue (perhaps encouraged by prizes etc.)
- the download of additional games
- links to game-based/theme-based fashion/retail websites.

Market Time Scales

With respect to the experiences taken from the Asian-Pacific market, where young people are extremely hungry for new, exciting services, it is believed that interactive games will be developed soon and introduced in all markets.

3.2.2 MOBILE SHOPPING

Description

Mobile shopping is part of mobile commerce. Mobile shopping groups together all services that allow you either to purchase and pay online for goods/information with your mobile phone, or to use your mobile phone as a payment terminal at public access points like drink machines or retailers.

Facilities

Mobile shopping will allow the customer to make:

- online purchases from traditional shops for delivery and online purchases from virtual shops for data and/or information (information, ticketing, music, video and games) delivery. This service will also allow users to have a multimedia (audio, pictures, video) description of the product they are going to purchase
- immediate purchase at retailers or public access points using mobile phone as a payment terminal.

Importance

On the Internet, mobile shopping is already a growing service. People, whether they are physically handicapped or whether they are so deeply involved in their daily business that they do not have enough time for shopping, do enjoy this service already today. They would like to have it available when on the move as well. Another point is the independence from having money physically available and using the SIM card in a similar way as people use their credit cards today for cashless payment in this environment.

Usage

User segmentation/profile: the targeted segment for mobile shopping is clearly the consumer segment.

Duration

The duration of each session will be quite low, around 1–4 minutes with a data bulk of some 10kb.

Number of Sessions

End-users are expected to access these services around once a day.

Benefits

Mobile shopping is growing on 2.5G and could be a key service for 3G with higher bandwidth and multimedia capabilities. Advantages for all actors in the value chain are important:

- for the end-user, ease of use
- for operators, a way to increase the ARPU and reduce churn

- for banks, a new and faster channel to market with lower operational cost
- for portals, content providers and retailers, a new personalised channel to market.

Expectations
All players, operators, content providers, retailers and banks expect a fast-growing 3G market when already established. This would be an excellent opportunity to offer mobile shopping to consumers. Consumers will expect reasonable prices for this service to compete with traditional high street shopping or Internet shopping today.

Drivers
- mass market mobile
- mobile commerce
- new technologies for wireless like WAP2, OS, GPRS, EDGE, UMTS/3G, Bluetooth and Biometry (i.e. fingerprint recognition).

Inhibitors
- secure transaction techniques
- standardisation (for encryption/decryption and digital signature)
- trusted third party
- shopping habits.

Source of Revenues
The mobile operator charges airtime to the customer to access the mobile shopping services.

Market Time Scales
Mobile shopping is available on 2.5G and will increase on 3G due to more broadband, greater speed and a growing market.

3.2.3 EXPERTS ON CALL

Description
Experts on call is an information advisory and consultant service where consumers and business people can seek and obtain specific expert advice at any time via their mobile Internet device.

An interesting consumer or business application is the ability to have experts of all types at your disposal instantly – possibly for the price of a telephone call (or less). Experts on call gives individual users access to instructions and information via voice and menu-driven information sources or live interactive exchanges with people who have expertise and specific information they can provide to the mobile user. A few examples of experts on call are:

- technical advice such as help making adjustments to a classic car engine
- the basics of how to change a tyre on an automobile
- assistance resolving everyday do-it-yourself projects or household appliance dilemmas
- first aid advice regarding a minor injury.

On some occasions the advice may even come from a close friend or a family member, because an agreement before a final buying decision saves time and money, whether it is a house, a piece of furniture or a gift for a friend, which corresponds with mobile shopping.

The phone/camera device virtually gives the expert the 'eyesight' he needs to advise you, help fix a problem, or make sure the right person arrives with the right tools and parts if a repair visit is necessary.

Importance

These important services are characterised by their ubiquity, speed and convenience of access. Setting up a connection from a UMTS handset would be in the reach of those who are not even computer literate. Such services have recently been made available to corporate users such as sales, marketing staff and managers. Some vertical markets use specific information data which would also be handy if accessible remotely. For instance, doctors would find it convenient to have access to their patient files while on a visit. It is anticipated that these services would be made available on a worldwide basis.

Usage

Peak data rates may be high for some types of services, but will be very occasional in general. Like today in the fixed desktop environment, it is anticipated that the mobile office extension will be used several times a day by employees outside of their office, whether on the corporate campus or on the road. Short and numerous sessions should result from frequently interrogating the corporate servers.

User Segmentation/Profile

The service is generally segmented into consumer and business users, with most people being interested in this service for potential assistance in various areas of their day-to-day life.

Duration

The session duration could be average to long, especially with interactive online sessions.

Number of Sessions

The average number of sessions will be low initially.

Benefits

In the same way that cellular phones were often bought for safety and security reasons with little intent to use the phone as a first choice voice telephone, likewise the experts on tap facility will move people into mobile multimedia, although initially these services will be of high value, but infrequently used. This type of service may become the norm for more routine customer service advice on buying, using or fixing products. In addition to benefits to end-users, this capability creates cost saving possibilities for existing customer services, as well as creating new earning opportunities for individual experts of all types, perhaps fronted by an agency or listing service, as you can find today on the Web.

Expectations

In some cases the financial or customer service costs saved by this service will greatly exceed the initial and usage costs.

Drivers

Since a camera is inherent in a mobile device, mobile phone/camera combinations and integration devices are likely to emerge over the next few years. As on UMTS a reasonable quality full colour image can reach its destination in under a second, the consumer and business applications for services are endless:

- extension of customer service offerings to include experts on call
- individual experts and agencies to provide directory and assignment.

There are various drivers for such services:

- today's way of life, where employees have less time and more things to do, requiring on-the-spot access and delivery of information
- penetration of mobile handsets into the corporate environment, bought initially for voice and increasingly enabled for data
- applications enabled for mobile data as shown by the initial success of data services in Japan and the recent enthusiasm for 2.5G services
- need to differentiate on service by companies in highly competitive markets. Such examples can be found in the overnight carrier delivery service, where real-time tracking and access of information have become key differentiating factors.

Inhibitors

There are a few inhibitors for these services:

- cost of service
- constraints in billing
- users do not become familiar because it is not frequently used
- coverage may not be ubiquitous
- mobile handset user interface where corporate information has been designed to be accessed via large screens and not small handset or PDA screens
- amount of data (bytes) required by the applications themselves initially designed to be running on the WLAN where bandwidth can rise from a few up to 54 Mbps per station (gross).

Source of Revenues

These services will generate several sources of revenues:

- data revenues for operators due to overall packet data exchanged between the employee handset and the corporate server
- licence revenues for software developers related to new technology and products
- service revenues for operators, integrators and software developers to enable the WLAN.

Market Time Scales

These services have already started on WLANs and will be deployed first as combined GPRS/WLAN/UMTS PCMCIA cards that have been introduced to the market already. It is anticipated that as UMTS is rolled out, there will be an immediate requirement to improve the bandwidth to get faster and more convenient access to information.

3.2.4 REMOTE MONITORING

Description
Remote monitoring allows users to manage different information at a distance, e.g. supervision of small children or your home when absent or while on holidays.

Importance
Remote monitoring will bridge physical distances. Compared to 'plain old telephone' remote monitoring will allow users to see and be seen whenever and wherever you are, additional to just listening and speaking. The evolution from radio broadcasting to television can give an idea of what you can expect.

Usage
It is difficult to say today how long an average session may last. On one hand, it may inspire people to use their mobile phones more extensively, because it may be similar to a physical meeting. On the other hand, the fact of having seen what the restaurant looks like, what the traffic situation really is or just a glimpse of a few minutes of daycare, customers may have received all necessary information and consequently the average session may take just a few seconds on up to a minute.

Number of Sessions
Many people would have at least a couple of situations every day where an image would give superior information to voice or text, starting with the traffic situation and today's lunch menu.

User Segmentation/Profile
With the broad range of applications, remote monitoring will offer huge value to all customer segments.

Benefits
First, the service would be beneficial for customers themselves. They would be able to watch or monitor people or assets any time and anywhere, an absolutely new service not available in other ways. That would give people a lot more safety and a better feeling when leaving home.

Second, companies offering remote monitoring may put up webcams at places of interests and make money on advertisements sold in connection to the broadcast. A site with webcams of famous cities like New York could probably be used to sell advertisements for hotels, restaurants and events in the city.

Expectations
'Tele-viewing' will be a great opportunity for customers to supervise people or locations of personal interest from any place in the world. It may give e.g. parents or holiday makers a better feeling and higher security when away from home.

Drivers
People would like to see or be seen instead of listening. It would allow flexible monitoring of people, places, property, environments or for health, safety and security. The range of sites to monitor is as rich as life itself. From the daily life perspective you can receive images of your

children while at school or daycare or you can monitor your house or car for protection and avoid traffic jams or long queues in your local stores. From a work perspective you can monitor construction sites and where people are waiting for your taxi service as well as queues in the canteen and delays of flights and deliveries. From a business perspective companies can provide their own live reporting of the wonderful ski resort they offer, the slopes, the view, the restaurants and the nightclub.

If the digital camera is part of a mobile device the usefulness expands to areas such as electronic postcards and experts on tap, but those are service groups valid on their own.

Inhibitors
The technology has existed for as long as the Internet and LAN but has limited acceptance as yet. Every form of surveillance without direct permission could violate personal integrity. Even with permission, like a video telephone call, users can feel uncomfortable (see Chapter 7). Legal issues are to be respected.

Source of Revenues
The revenues for remote monitoring will be triggered by digital camera equipped mobile phones, which would make carrying additional digital cameras obsolete. The camera quality should be similar to traditional digital cameras (resolution better than 1 megapixel). The subscriber might have to pay to see the latest traffic situation or a scenic view while the local bar or ski resort probably would take it on their marketing budget to show live images from their establishment. This scenario is already intensively advertised by network operators and includes videoclips of some 10 seconds as well.

Market Time Scales
Camera and videophones as well as webcams already exist. Given the low cost of camera phones, every tourist attraction and public location could easily be covered. The technology to transfer images in a compressed format is also well advanced. With 3G terminals available generally equipped with digital cameras everybody has his/her own 'TV monitor' in their pocket.

Integrity will be a major inhibitor. It is against the law in many countries to put up surveillance cameras without permission. This limits the spread to situations where the viewers and those being monitored have some kind of agreement, like in the nursery. Some bans have already been imposed and may be a barrier in the evolution of 3G (see Section 2.4.3.2).

3.2.5 TELEMATICS/TELEMETRY/MONITORING

Last but not least there is another field of services, the tele-services area. A specific example of this class of service within the context of UMTS (3G) telematics includes services such as self-diagnosis checks for lorries/trucks and cars before breakdowns occur, the provision of a breakdown service when the vehicle has an immediate fault, handling emergency calls when the car breaks down, and positioning information giving the exact location of the car. Several big manufacturers in the European automotive industry are working together to develop revenue-generating content services for the 'intelligent car'. Some of them have already incorporated telematics into their premium-class vehicles, primarily to monitor potential emergency situations, where the additional expense can be easily absorbed. Next to the mobile

Table 3.2 Telematics and telemetry specific applications

Segmentation	Service	Benefits	Example
Alarm and security	Commercial/residential security alarms Smoke/fire detectors	Alarm and status messages sent to alarm service centre Security against compromise of wire-line connections	Burglar breaks into a house and cuts the phone and power wires Alarm company is notified of break-in and dispatches police
Agricultural, irrigation and environmental	Pipeline corrosion monitoring systems Water pump failures, levels of contamination Air quality systems	Monitor environmental condition Alarm systems for hazardous environmental conditions	A city registers unusually high air pollution readings Public service announcement is sent out warning those with medical conditions to stay indoors
Asset management and tracking	Office equipment Industrial machinery and manufacturing processes Vending machines	Monitor meter information Service diagnosis and maintenance Inventory management Fleet/route management	A delivery truck follows a specific route to fill drinks vending machines Truck can be re-routed if the vending machine is still full from last delivery
Atmosphere controls	Heating, ventilation, and air conditioning Refrigeration, temperature and humidity controls	Air quality standards Temperature controls for food and other perishable items Climate control for greenhouses and agricultural products	A commercial refrigerator door in a restaurant is left open after a food delivery An alarm is triggered, alerting people to close the door to save the food from spoiling
Public and municipal services	Parking meters Highway tolls Railroad crossing Switches	Parking meter servicing Service route management	A city is losing money everyday because parking meters are full or robbed
Transportation systems and facilities	Vehicle location Vehicle engine computers Container asset tracking	Inventory management Emergency communications Vehicle maintenance Navigation services	A truck is making an urgent delivery but there is a major accident that delays arrival; information about alternative routes
Utilities	Utility meter reading Oil and gas pipeline facilities	Reduced cost of servicing meters and remote meter control Customer account management and additional enhanced cost saving services	A customer is moving out of town Utility company can provide up-to-the-minute billing to settle the account

© UMTS Forum (2000b).

equipment is the in-car navigation system, which includes real-time mapping and street-by-street guidance services.

Table 3.2 illustrates the wide range of specific services for telematics and telemetry in the field of UMTS (3G) systems.

3.2.6 ADVANTAGES FOR MOBILE WORKERS

More and more people are working as mobile workers. They need access to their office files and documents from many different places, wherever they are and whenever there is a necessity or an opportunity, whether it is e.g. at conferences, on the train, in an airport lounge, with the customer etc. You do see two different types of mobile workers, those who need occasional access to bulky files, graphics, pictures etc. and those who are more nomadic and need access anywhere and any time. The latter are on the road almost day by day and will have frequent access. The difference between these two mobile worker types was discussed in deeper detail in Section 2.4.1.2.

By provisioning more bandwidth and enabling data mobility between the corporate user and the central enterprise information centre, UMTS will enable a range of mobile services which until now have only been available to users through fixed desktop clients. Such services will typically include email, intranet, synchronisation and directory access.

In an era where the bulk of corporate information is available and accessible anywhere and any time, the immediate access to this information throughout any telecommunication network, whether it is via a fixed-line network, like public switched telephone network (PSTN), integrated services digital network (ISDN) or digital subscriber line (DSL), or satellite or any kind of wireless network, will minimise time loss and improve the employee's productivity, synchronise all the company's resources and enhance customer care.

3.3 PORTALS AND THEIR BUSINESS MODELS

Leading mobile operators recognise that their traditional wholesale approach to corporate and residential customers restricts their opportunities. A 'pipe-only' wholesale strategy means the operator simply connects the customer with the existing Internet. On the business side, providing only access and transport means:

- operators can compete merely on the basis of transport prices;
- revenue streams are limited to access and transport capabilities;
- with number portability customers can increasingly substitute one carrier for another;
- service offerings carry the brand(s) of other providers and competitors;
- other providers 'own' the customer by controlling the end-user profile and records.

Figure 3.2 illustrates the relationships between all involved parties. The figure shows that the user's relationship is mainly between his/her service broker/management (service company = ServCo) and the network operator (network company = NetCo). In general, the providers do not have direct access to the subscriber, although they may have the end-user's profile and records. Within the user's home environment (HE) the service broker enables the customer to act in 'one-stop-shopping'. The advantage for the customer is that he/she has only one

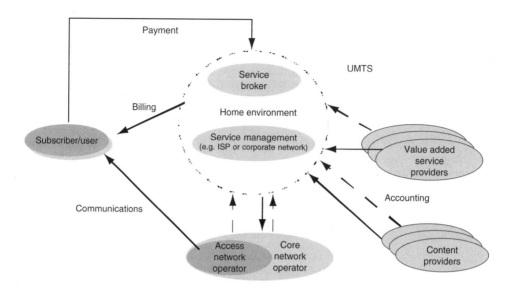

Figure 3.2 Service management and business relationships between all involved parties. © UMTS Forum, report no. 2, 1998.

contact point (the broker), who takes care of all back-up services. Quite often, the service broker represents the network operator as well or vice versa. So, they deliver the portal services to the end-user.

3.3.1 DEFINING A PORTAL

The current popularity of portals arises directly from the information explosion. Hence, finding information becomes very difficult and even easily getting to it is more difficult. The definition of a portal has to do with virtual marketplaces within the internet and intranets and a single point of access to aggregated information – some place you go through to get something else. It could be a Web 'supersite' or search engine that provides a variety of services, including Web searching, news, white and yellow pages directories, free email, discussion groups, online shopping and links to other sites.

A search engine is the basis, which works like a navigator for websites, but in the future also for direct marketing and e-commerce.

Search engines enable portal operators to control Internet commerce; together with servers owned by an Internet service provider (ISP), they also control end-user access and take over administration. Very important is the capability to optimise the search process according to the user's profile and business objectives of the portal operator. For further description, a portal is defined as follows:

A portal is an entry point to a wealth of information and value-added services. Portals can be personalised and are Internet/intranet-based with browser user interfaces. Portals will deliver content according to the device's characteristics and user's needs.

A portal is not something that is set in concrete for eternity; in a fast-changing business world a portal is something that changes with the requirements. Thus, a 'portal roadmap' may show the evolution in this area.

A portal offers:

- relevance and response
- personalisation
- value-added services.

Users go to a destination portal and find what they need quickly. Portals provide personalised information in the format users really want, and users get addicted to services such as email and instant messaging, speech and audio/video etc. These points are valid for Internet operators as well as for cellular operators who build a 'portal platform'. To be successful they should know their customer needs and requirements and consequently provide them with remarkable value. The most critical metric is how many unique subscribers a portal has who will return to the site.

Three trends of portal developments are visible today:

1. Content-oriented portals
2. End-user-oriented portals
3. Convergence-oriented portals.

Content-oriented Portals are tailored to search in catalogues and towards specific market segments, e.g. Infoseek, Amazon, TV portals. End-user-oriented portals are specialised for certain user categories, e.g. wireline versus wireless portals, intranet portals, WAP (mobile) portals. The concept of convergence-oriented portals is the integration of various user categories via different access networks into a multi-portal, e.g. the multi-access portal *VIZZAVI* (a European cooperation project of Vodafone, V-NET, SFR, CANAL Plus, D2, Omnitel and Telepin) or *t-zones* (T-Mobile), providing multi-device services or multi-linguality. These portals do include content management for data, speech, messaging, audio/video, further user-related data based on users' profile and location, unified mail and accounting.

A UMTS operator needs to consider whether to simply provide a wireless Internet protocol (IP) pipe to a service offering hosted elsewhere on the Internet, e.g. at a portal like Yahoo!, AOL, T-Online, Excite or Infoseek, or to invest in building value in their own brand and end-user relationships by offering differentiated services hosted ex-house or in-house. The wireless IP pipe business using tunnelling will tend towards a commodity bit-shifting operation, where cost, coverage and data rate are the only competitive dimensions. By carefully developing and preselecting useful Internet-based mobility services with careful branding and competitive tariffs, operators can encourage the user to buy into their propositions.

Convergence between the mobile telecom industry and the Internet and media industries will play an important role when these industries are positioning themselves for the future revenues of wireless Internet. This has large implications for the mobile network operators' ways of doing business, and eventually they will have to redefine their position in the value chain. By taking the lead in wireless Internet development, network operators can retain their strong position in the future wireless industry.

3.3.2 KEY SUCCESS FACTORS

There are two major groups interested in portals: those who provide network access (network operators) and those who execute transactions. For both groups, the same factors – reach, richness and affiliation – are important when it comes to the selection of the portal. Network operators can be very strong in at least two of these dimensions – reach and affiliation – and are therefore well positioned in the portal industry. The third dimension – richness – can be covered by alliances with content aggregators.

Reach

For the portal user, the value of the portal will be (over-)proportional to the amount of information that the portal provides. Furthermore, the number of other users that the user can be connected to is important as well. However, it is not only the number that counts, but the quality of segmentation of this potential market is also an important attribute of an attractive portal.

Richness

Within a sufficient reach to potential customers and to service offerings, the richness and quality of the information offered is important. The more and the deeper the information on a certain item a site can deliver, the higher the potential value to the user will be. This is true not only from the user perspective but from the site's (company's) perspective as well. For them, a high richness of information about the users translates directly into segmentation quality. Portals that attract well-targeted consumer groups, because they are based on a fine-grain need-based segmentation, are of much higher relevance to the business community than the very broad general-purpose portals that exist today.

Affiliation

For the user of a portal, it is important that the portal provider is affiliated to the user rather than to the sites the portal gives access to. To rely on the information provided by the portal, the user must be convinced that the portal provider acts at least in a fully neutral way or, even better, in favour of the consumer. It is highly likely that portals that are credible in communicating their consumer focus will be much more successful than business-minded portals.

3.3.3 PORTAL AND CONTENT

Providing the portal does not imply creating the content. Content feeds are likely to come from existing content providers or other portals, but with subscriber data being captured by the operator. It naturally follows that the first applications should focus on building the subscriber base and increasing airtime, namely messaging applications. Next should come applications which drive subscriber profile data capture, such as personalised subscription-based content push and wireless personal information synchronisation. Only after building the profile database can mobile e-commerce and advertising be successful.

For operators with strong brands and a desire to reduce churn and increase their share of the customer value chain, becoming a true mobile ISP means offering their own mobile Internet services to users on a wireless portal platform integrated with billing, customer care and positioning systems. Typical examples are T-Mobile's *t-zones* and Vodafone's *Vodafone Live!*.

3.3.4 THE IMPORTANT ROLE OF VIDEO IN THE CONTEXT OF PORTALS

Video will become a dominant medium on the Web. Thus, video portals will become important. Moreover, portals give access to text-based information without understanding the markup language and its meaning. Understanding the markup language requires understanding of the semantics. Video portals will be designed with shallow semantics to organize video data. This embraces search and retrieval within an environment with video facilities, offered through suitable communication channels (DVB etc.) The so-called 'Vortal' (vertical portal) will focus on a particular subject, e.g. books, CDs etc.

To accommodate the continuous enhancements by a phased approach, a mobile multimedia portal platform is defined, which enables the UMTS operator to deal with the new market segments.

Exclusive agreements between content providers and content carriers as a result of convergence may be seen to limit consumer choice by excluding access to content provided by competitors, especially if there is insufficient effective competition in the provision of delivery channels to the user. Possession of rights to key content, such as major sporting events, may give market players particular commercial power, and this is where the application of competition rules will need to prevent abuse of dominant positions.

Alternatively, convergence may have the effect of dissolving access bottlenecks. For example, the exclusive distribution rights awarded to cable television companies may no longer lead to monopoly power at service level. Cable companies are likely to compete with IMT-2000/UMTS operators, digital satellite and terrestrial television broadcasters and Internet access providers.

3.3.5 POSITIONING OF NETWORK OPERATORS WITHIN THE PORTAL INDUSTRY

The UMTS operator generally has, compared to other potential portal providers, a very large customer base of several million customers. In addition, it has a very clear picture on the segmentation of the customer base. By virtue of this customer base, the operator is set to excel in the dimension 'reach' both for its own portal users, who can communicate with a large number of other portal users, and the business community as a whole which is offered access to this customer base.

Therefore, the operator portal is likely to be attractive to a large number of companies as well, increasing the value of reach for the portal users. This is setting in motion a positively self-reinforcing cycle that will ultimately provide high value from reach to both users and providers of sites.

The richness of information provided by the various portal sites cannot really be influenced by the portal provider (the operator). However, by offering a very valuable user base to the portal sites, the operator is attractive as an alliance partner to content aggregators that are very able on the richness dimension. Since the number of premium-class aggregators is limited, alliances of operators with a large customer base and aggregators with high richness of information look like a sure-win business alliance.

3.3.6 THE ROLE OF AN INTERNET SERVICE PROVIDER

The simple definition of an ISP is providing access to the Internet. The level of service, the set of functions supporting the end-user from the ISP, varies. The client–server model leads to the 'user@domain' account (point of presence, PoP) which means that the ISP has to manage the user's addresses and applications-related protocols. The ISP's services can range from tunnelling (see Section 4.5.1.2) services up to mail or http hosting services etc. In all cases of accessing the Internet via tunnelling, http, email or FTP the ISP has the responsibility for the end-user's account. As a consequence the 'all-IP' UMTS network takes the UMTS operator into the position of an ISP. Combined with the extended functionality involved in dealing with the roaming user, the role becomes that of a mobile ISP. The additional complexity comes from inter-operator roaming.

3.3.7 THE ROLE OF THE CONTENT PROVIDER

The content provider has an increasingly important role, not only as a simple provider of content but also by adding value (data mining, repackaging etc.) or even offering complementary if not competing services (owning or associating themselves with portals). In traditional media, content providers are owners of content, e.g. broadcasters, publishers or people who aggregate content created by third parties, e.g. music/film producers or authors. They have no mobile experience, no idea of user location, and no billing capabilities for combining different kinds of content or services.

In any case, no one is better placed than the content provider to provide content in the appropriate format – except that it has to get such requests via the operator, in many cases as a function of the network conditions. On the other hand, more and more data will be available as meta-data, the same content delivered in a variety of formats. Then, only in a few cases will there be provisions for transcoding the information, and that can be done either at the content provider or the operator side.

3.3.8 REQUIREMENTS FOR PORTAL SET-UP

The role of a portal with regard to the presentation of information can be perceived in two perspectives: that of the end-users and from the primary attributes of a portal owner. However, as the notion of what is the core business of a company evolves in time, portals, especially if successful, will continue to spin-off.

3.3.8.1 The End-User's Perspective

The end-user of a device using such a portal to obtain access to the information and value added services available will have certain expectations.

Presentation of Information

A user expects information to be presented in a format appropriate to the device being used. Since one major goal of a portal is to induce the user to use the portal as the focal point for access to information and a user may use more than one device to access the portal, the portal (application server) needs to be able to determine the device being used. While a simple list of terminals commonly used could be presented, this is not advocated. A better implementation would use an automatic means of identifying the device and prepare (i.e. transcode) the information in a way that best suits the device characteristics. Techniques to discover the device's capabilities are being specified by the World Wide Web Consortium (W3C) as CC/PP[2] and by the WAP Forum as User Agent Profiles.

A device will in many, if not most, of the currently envisaged cases use a Web browser as the means to present information to the user. However, this may not be the only application model, and the use of Java and other techniques that provide for device-independent authoring is envisaged, with the device determining the exact means of presentation. The WAP Forum's device-agnostic authoring model using WML is one simple example.

The use of speech as an alternative means of delivery of information may also be envisaged. The accessibility initiative of the W3C and ongoing developments in the area of Design for All are good examples of the approach required to meet the presentation needs of handicapped users or simply of those who prefer or need something other than the presentation of textural and graphical information on a screen.

The user may also be interested in the portal providing access to a unified messaging centre, thereby allowing access to SMS, email, faxes, voicemail etc. through a single preferred user interface. And finally, the user may be interested in subscribing to particular user groups/communities and having access to all relevant information.

Best Quality of Delivery

The user will expect the portal to be capable of delivering services that allow the capabilities of the device to be exploited to the fullest, e.g. highest definition of graphics, rich text, high performance audio. However, the user will also expect to be able to set preferences backing off from this best quality case (e.g. lower definition graphics or simple text, reduced quality audio), in the first instance and to choose selectively a higher level of presentation quality when required. This is particularly true when using wireless communications, where quality usually translates in terms of both delay and cost. Depending on the specific system being used and upon the network load, the user profile should specify the maximum allowed data rate (for bandwidth-on-demand systems), allowing this default setting to be overwritten by the user.

Portal Quality of Service

A user will have a minimum QoS expectation from the portal. This goes well beyond simple availability of the service and may include aspects of performance and how the service gently degrades as link performance and system loading affect the underlying wireless network performance. Quality of service can also be considered a cost issue to the user. A portal and the associated wireless link management should be able to offer the user selectable levels of delivered quality and performance based on link bandwidth (i.e. cost), system loading (which varies with the time of day) etc., and thus should be able to measure these factors.

---

[2] The Composite Capability/Preference Profile (CC/PP) is a collection of the capabilities and preferences associated with the user and the agents used by the user to access the World Wide Web.

Roaming

Roaming can also present additional problems. Depending on the topology and cost structures of accessing the home portal service when visiting another wireless network, the user may require a change of settings to manage costs and performance. Some current implementations will not allow access to the full portal functionalities from the Internet and sometimes rely upon agreements with other networks to allow access to the home network via secure links from local numbers. The user, however, would expect access to all information regardless of where he/she is.

Customer Care

Personalised Service

Users will expect to be able to select their preferred banking, stock trading, multimedia video and audio services, and m-commerce sites to supplement a set of basic services associated with the subscription profile.

Such subscription profiles may well contain a basic set consisting of information, news, weather, email, and access to voice and unified messaging services. This basic profile may well be supplemented by additional choices reflecting additional subscription or corporate services. Finally, there are the 'pick-from' services, where banking and e-trading are good examples. A user cannot be expected to use the bank or e-trading company the operator of the portal prefers. Thus the portal owner allows users to select their preferred services from a list or even to add their own entries.

Finally, the amount of personalisation may depend on the user subscription. Free portal services or those linked via the portal may well be encumbered with banners and adverts as part of the business model. Such overhead penalises the user in terms of performance and potentially additional costs if flat rate billing is not being provided. If the user pays a subscription for the service, it is a reasonable expectation that access is completely personalised, even to the extent of having banners and adverts removed should this be desired.

Privacy, Trust and Security

Users have an expectation that the portal will respect their privacy. Unless explicitly authorised, users will expect information collected by the portal not to be shared with service providers or any other organisations who would like to use such information for e.g. targeted advertising etc.

The user also has an expectation of trust in the portal. Without trust the user is unlikely to make this the focal point of their access to information and services. The portal needs to meet this expectation of trust by providing reliable information, a secure and trusted environment for m-commerce, reliable billing, respecting privacy etc.

Support

The user can expect support from a portal services provider. While portals on the Internet have relatively low levels of support, often available only via email, this is tolerated because the cost of using such portals is very low, and in many cases zero as ISPs are subject to ever-increasing competition. Where a user is mobile and has few alternative means of requesting support, they can expect a higher level of immediacy and contact with the customer service of the portal owner. The cost of this support may be included in the subscription or may be separate, depending on the type of subscription or problem.

For devices with sufficient capability, the portal services provider or application developer might wish to offer additional client software to supplement the basic browser etc. used to present information. The portal should be capable of meeting the users' expectations should such devices become available. Further maintenance of such applications, and potentially even of the device's firmware, can reasonably be expected to be supplied via access to the portal in the same way Web access is used to provide and maintain applications on the personal computer (PC) and personal digital assistant (PDA) devices in use today.

Billing

The user already has a commercial relationship with an operator or services provider. Using a portal service is likely to imply additional service charges. The user is likely to demand consolidated billing from the portal owner for all services provided via the portal. The exception might be higher value purchases where direct billing to the user may be the only appropriate means of payment.

However, large bills are always problematic. From the user's perspective, large bills often result in mistrust or scepticism and from the operator's or services provider's perspective often resulting in angry queries and higher customer service costs.

From other services, e.g. television, the user builds a perception of acceptable costs. A service may be free when encumbered with advertisements, while such adverts would be unwelcome if a service is paid for via subscription. The user will certainly expect a clear indication of incurred costs of access and service usage from the portal and be able to select how much advertising is tolerable, and from which sources, in lieu of access charges.

3.3.8.2 Portal Owner's Perspective

Formatting to Meet User and Device Requirements

The portal must meet the need to provide content suitable for the device and the user's preferences. While the technology exists to automatically adapt the content to suit every device, the portal owner may wish to group different devices or device attribute capabilities to reduce the number of permutations. In so doing, a device needs to be related to the best-fit group when the profile for that device (user agent profile, UAPROF) or set of attributes (CC/PP) is disclosed during session establishment.

Customer Care

Arguably, the portal needs to support a variety of profiles for an evolution of basic subscription service bundles and to cater for various virtual private networks (VPN) and corporate intranet extensions. Further, the portal needs to allow personalisation to meet the needs of users, e.g. choice of banking service, stock broker, m-commerce etc.

The portal also needs to define typical templates to help users define the look and feel of their personalised portal within the constraints imposed by the portal owner. These templates will obviously allow for the inclusion of services but also of additional information. For example, a user who is worried about the cost of services or transactions might be able to add the current bill or session cost indication to the screen, in the same way a word processor can add the page numbers and keep them up to date.

Finally, a well-designed portal will support both 'pull' and 'push' models of operation for information services. For example, telematics and stock price changes can be profiled to inform the user when traffic is bad en route during a planned journey or when the price of a stock in the user's portfolio moves $+/-x\%$ to allow appropriate action to be taken. A 'pull' model might be very well suited to, e.g., white/yellow pages directory lookup.

Privacy and Trust

In a service such as a portal where the user is being offered a personal service an enduring relationship with the user will only be achieved if the user trusts the portal. Trust is a complex feeling but two important aspects are security and privacy.

A user needs to feel that the portal respects his/her rights to privacy. Where communities of users are facilitated, for e.g. chat or even sharing personal interests, it is important to ensure the user can control the subscription to such communities with confidence.

Whether as part of a community or as an individual, protecting a user's access to portal services, the user's identity and other personal information (e.g. PIN, address, credit card information) is a key requirement. Only such information as has been explicitly authorised by a user should be divulged to other users or groups.

Security

Several aspects of security affect trust. Just some are included here. The security of any information being exchanged between user and portal is important, and techniques such as encryption of the transport link can be used to achieve this. Ensuring an accurate consolidated bill of services and m-commerce transactions is critical, and ensuring adequate authentication and non-repudiation are key techniques to achieve this.

Finally, it is important to ensure that the appropriate security mechanisms are applied to various types of service offerings, e.g. pay-per-view, subscription and free content business cases. A user may be more concerned about the data received and the accuracy of billing for a pay-per-view or transactional service offering and might be particularly offended if unsolicited adverts are received during such a premium service.

Monitoring Users' Behaviour (Heuristics) and Legal Aspects

While a user may be able to personalise a portal to meet his/her individual requirements and preferences, the use of heuristic technology can help. The gathering of usage logs and application of data mining technology can enable heuristic behaviour capture, providing the user with updated services and suggestions, based on recent past usage, for new services to be added to the personal profile. Eliminating old services, even those seldom used should be left to the user.

Where lookahead technology is being used to increase apparent performance of slower bearers, such heuristics can be used to modify the content for a user to include most likely next locations and to pre-cache, awaiting the most likely next request.

Legal requirements are placed on ISPs and other service providers, especially those holding personal information about users. A portal owner must ensure that such a service meets all these requirements. This might well entail having to log a user's use of the portal beyond the level required for billing and associated customer support. Issues of legal interception fall into this category.

Support

User need constant support, whether it is related to troubleshooting on their system via hotline services or qualitatively on portal access or content items. Services providers must take note on this important issue of customer care, otherwise they may soon lose their clients, who will look for better support from other more service-oriented providers.

Billing

At issue is the location of the subscriber profile records, which reflect the personalised service choices of the end-user: message filtering options, choice of mobile information and type of mobile device, correlated with name, billing address, mobile phone number and email address. This store of data will permit additional returns through selectively targeted mobile e-commerce and advertising. So, can wireless network operators resist Internet companies' attempts to redefine the value relationship to end-users in their favour? UMTS operators have three distinct advantages:

1. They send the subscriber a bill fixed as quarterly, bimonthly or monthly and charge the subscriber on a time-dependent call basis. This makes it possible to get a revenue stream via a small additional charge for mobile Internet services and means that a business return is possible well before mobile commerce and advertising become feasible. Added to traffic revenue, the operator is in a commanding position to capture or selectively share this revenue with value chain partners on its own terms.
2. The operator provides packet transport with the mobile service, necessary to integrate Internet services with intelligent network, voice, data and fax services and allowing volume-based charging in addition to time.
3. The Operator will know the subscriber's location using emerging cellular positioning technologies. This is high-value information that should not be given away to other value chain participants. Positioning adds end-user value through information customisation, e.g. details of the nearest restaurant or automatic conversion of email to speech for a driver of a moving car. Location information will enormously increase the revenues, combined with content, an item which was already mentioned in Section 3.1.1. This will allow applications-related charging.

With these three advantages, the UMTS network operator has a sustainable position in the value chain and a unique ability to deliver certain end-user applications such as integrated media conversion for messaging, location-sensitive push/pull content applications, and micro-browser control of telephony features such as call diverting. By carefully tailoring the value proposition, operators can reduce churn and avoid becoming commodity bit-shifters. Operators will need a suitable, scaleable, mobile Internet portal services platform on which to offer such services with integrated billing.

Provisioning and Branding

The use of a portal often results in changes to the personalised list of services the user intends to call upon in the future. This needs to be reflected in the provisioned personalised service, and so a close linkage between the provisioning system and the portal needs to be considered.

Users choose which categories of the bundled or aggregated services they want to have in their created profile list. Each user will have a unique set of choices. The portal will have to monitor and dynamically update each user's choice. In addition, the appropriate tools will have to be made available.

Since people use a variety of devices for input and output, user agent developers must ensure redundancy in the user interface. Messages and alerts to the user must not rely on auditory or graphical cues alone; text, beeps, flashes and other techniques used together will make these alerts accessible. Text messages are generally accessible since they may be used by people with graphical displays, speech synthesizers, or Braille displays.

The following requirements should be provided:

- Ensure that the user can select preferred styles (colours, text size, synthesised speech characteristics etc.) from choices offered by the user agent. The user must be able to over-ride author-specified styles and user agent defaults.
- Ensure that the user may turn off rendering or stop behaviour specified by the author that may reduce accessibility by obscuring content or disorienting the user.
- Ensure that users have access to all content, notably author-supplied equivalent alternatives for content such as text equivalents and auditory descriptions.
- Ensure that the user can interact with the user agent (and the content it renders) through all of the input and output application programme interfaces (API) used by the user agent.
- Provide information to the user about content structure and metadata to help the user understand browsing context.
- Alert users, in an output-device-independent fashion, of changes to content or viewports, but also to recognise the best features and the most used ones.
- Manage tools to measure performance and to bundle/aggregate services and repair Web content, as well for new tickers, applets and new windows to create profile and content.

The goal of a portal owner (e.g. an operator) is to build the relationship with the users. A major step is to create a strong brand awareness for the service. Factors such as banners, logos and presentation style are the classic ways to build this awareness of the service's brand but more important in the context of a portal is the provision of appropriate information in an intuitive and well-organised way.

3.4 SERVICE PROVIDERS' BUSINESS STRATEGIES

When discussing potential business relationships, various models can be considered, comprising the roles of the network operator, the Internet service provider, the portal operator and the content provider.

- Fragmented model: the roles remain separated. In order to satisfy the end-user, a certain degree of harmonisation and control between the independent functions will be difficult to fulfil.
- Partnership model: the main roles (network operation, ISP, portal) are maintained in cooperation. Cooperation agreements will be necessary for acceptable service offerings to the end-user.
- Ownership model: it comprises the main roles in one ownership. Clear responsibility lies with the operator in providing end-user services.

Of course, the ownership model has to allow additional partnership and/or additional service offerings from independent providers, e.g. end-user access to other portal operators and other content providers. For a better understanding, the following description considers two alternatives through which the user can access the content.

Scenario 1: Fragmented Model
Figure 3.3 depicts the scenario where all the roles are kept completely separate. Attention is now drawn to the fact that, in this condition, all the subscription and security data related to all the involved operators/providers should be stored on the UMTS Subscriber Identity Module (USIM) in order to ease the user's behaviour. Agreements would then be needed between the companies, which play different and separated roles in the business chain.

The user needs to register with different companies – a UMTS network operator, an Internet service provider, a portal provider and a content provider – to be able to access content. In this situation, the Internet becomes the most likely interconnection network between the different operator/provider domains. This means that, on one hand, the parties involved in this path can choose a completely different way of handling mobility, QoS and security. On the other hand, one important segment of the transmission connection, i.e. the Internet domain, cannot be controlled at all, being outside the reach of all the operators/ providers. In the case of location-dependent services, forwarding the user location information underlies technical and legal conventions.

Scenario 2: Ownership Model
Figure 3.4 represents the situation where the UMTS network operator also plays the roles of service, portal and probably also content provider. This means that this path is completely controlled by the UMTS operator, both in the UMTS system and within the operator IP backbone. The operator can automatically decide which solutions for mobility, QoS and security

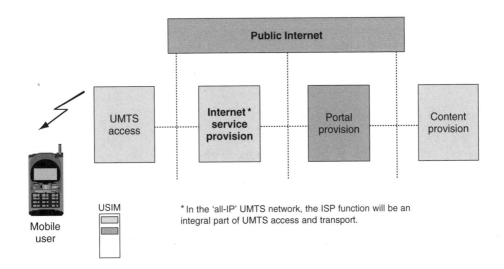

Figure 3.3   UMTS access and other service providers separated. © UMTS Forum.

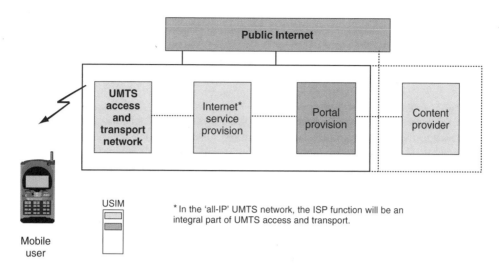

Figure 3.4 UMTS fully fledged ISP/portal operator. © UMTS Forum.

control are the most suited to its business model, since all the nodes and networks are under its own control. In particular, the handling of mobility, QoS and security issues is eased by the fact that all the adopted solutions are known and designed to work together. The UMTS operator can also provide the user location information under its responsibility to its own applications and content offerings.

All the related subscription and security data on the USIM module belongs to the same legal entity. It is quite important to note that a particular content could be money, when the correspondent content provider is a bank. Then this example shows a scenario where the UMTS operator can also assume the role of a bank, where the money stored in the USIM can be used to purchase products or services. In this way the USIM becomes an electronic wallet controlled by the UMTS operator. Optionally, the user can, even in this case, have relationships with an independent Internet service provider, possibly for a different set of services not provided by the UMTS operator. However, the data related to this provider can also be stored on the same USIM.

3.4.1 THE BILLING CHALLENGES AND OPPORTUNITIES

The move away from telecommunications towards information delivered via IP infrastructure presents an entirely new set of challenges for the billing manager, the pricing manager and the business as a whole. The first and most fundamental question that arises from the advent of IP services is: 'what will we bill?'.

The traditional elements that were billed in the world of telecommunications become irrelevant. Time-dependent billing on the network is already fading and will disappear quickly as customers get used to the idea that connections are 'always on'. Distance will disappear quickly, too – IP addresses are always 'local'. The answer then becomes 'data' or 'volume'. The cost of transporting information must be covered and, at least, cover the cost of building, maintaining, extending and eventually renewing the network.

However, the real opportunity for service providers will be to take their place in the value chain of m-commerce, and in return for billing and supporting the customers receive a percentage of revenues of the products and services bought. In addition, there is the opportunity to act as a wholesale 'agent' for content providers, selling on products and services, with a mark-up.

Bearing in mind that the service provider and the access/transport provider may be the same corporate entity, it makes sense for this entity to own content as well, in order to maximise revenues from these new services. In addition, third-party content can always be delivered to customers.

There will be no single business model, but a number of different unique models. Each agreement between service provider, access/transport provider and content provider could be different. The key to an efficient billing process will be flexibility. The key to an effective billing system will be scalability, as the volume of billing data being filtered by mediation devices increases from today's levels by factors of five or ten.

3.4.1.1 Billing for Access/Transport

While it is generally agreed that 'transport' must be accounted for, it is also agreed that the margins on 'transport' will continue to decrease. However, for certain services, the QoS at which services are delivered will be of recognisable value to the user. As examples, a user will require a certain QoS if watching a video, listening to music or taking part in a video-conference. Different levels and parameters of QoS will be required for each service.

At a billing level, QoS parameters are well understood. However, a dialogue needs to be formalised between the billing system vendors, mediation vendors and network equipment manufacturers in order to ensure that the relevant information can be passed from network device to billing engine in a timely fashion.

3.4.1.2 Billing for Corporates

It is likely that the complexities involved at the consumer level will be simplified to some extent when offering service to corporates. Companies are familiar with the concepts of 'data' and 'bandwidth' and are used to negotiating with their service providers for access and use of high-bandwidth services. A wide range of information services will be catered for in this way.

However, complications will arise when executives are using UMTS services when travelling, particularly internationally.

3.4.1.3 Billing for Content

As discussed elsewhere, 'content' can vary from basic information services, such as stock exchange information, to downloadable or streamed video, to transactions taking place over UMTS networks.

In order for Service providers to gain the potential to bill for these transactions and services, the first step must be for them to be able to identify the content that is being delivered. At present it is difficult to achieve this, and service providers such as ISPs are not able to take advantage of the full potential of e-commerce.

The Internet Detail Protocol Record (IDPR) organisation, based in the USA, is currently working on a standard for identifying usage and content on IP networks. The format has to be open and flexible. The current (iterative) approach is to define essential parameters for any IP transaction and to provide a mechanism for extension, which enables efficient communication between service elements, mediation layer and business support systems. IPDR release 1.1 is now available, and activities involving 'proof of concept' are in progress.

3.4.1.4 Billing Traditional Users

Currently, terms such as 'always on' and 'data' are not widely understood by the user. Therefore, if the services and products that can be accessed and bought via UMTS technology are as near to the real world as possible, then the service acceptance by the user is likely to be much faster. This would be much more understandable if he/she were to buy a video or a piece of music than a 'megabyte' or QoS.

There will certainly be many different types of services to collect money for, and these may be a mixture of access fee (perhaps subsidised by advertising, if requested), subscriptions to various services, usage of the network and transactions. It will be further complicated by the fact that a user will do several things simultaneously during a UMTS session.

The user, both now and in the future, will require clarity and simplicity from the bill. Service providers will need to think carefully about how much detail to put on the bill of the future, particularly as bills will begin to resemble credit card bills. It may be that a certain amount of 'transport' will be bundled as part of the service, and only charged for as line items once certain thresholds of usage have been reached. QoS discounts will need to be thought through, in terms of presentation.

The gaining acceptance of electronic bill presentation and payment (EBPP) will play a major role, not only in the context of the virtual home environment (VHE), and will be a key factor in attracting customers and, indeed, keeping them (customers are less likely to churn if using multiple services and/or EBPP). Presenting bill summaries via UMTS will be an attractive feature. Indeed, presenting bill summaries for a wide range of services will inevitably attract users. It incorporates the need for simplicity with a mechanism for presenting the customer with as much detail as required, without 'pushing' all the details to the customer in the first place.

In addition to user or retail billing, areas of complexity will arise when reconciling or billing content providers and other partners for services delivered or bought as wholesale items.

3.4.1.5 Roaming Impacts

As well as being an essential requirement for regular travellers and an important feature for infrequent travellers, roaming is a very important business issue. In the ideal UMTS world, users would be able to use local services wherever they are. Service value will be an issue as services need to be agreed in roaming contracts and additional services have to be added in the charging record format that is exchanged between operators. For example, in the GSM world this means new service type(s) in transfer account procedure (TAP) format, which will probably also be used in UMTS networks. The prepaid concept will add additional complexity.

3.4.2 PRIVACY, DATA PROTECTION AND LEGAL INTERCEPTION

In order for convergent services to develop, users need to be assured that their privacy is adequately protected and, in particular, to have confidence in the security of information passed over the networks they use. Where location-based services are used, appropriate safeguards for personal data and privacy protection need to be implemented.

An agreement on privacy and data protection has been signed between the USA and the EU, but many other countries either have a limited privacy and data protection regime in place or do not have it at all. Legislation has already been agreed at the European level in the form of legally binding Directives.

3.4.2.1 Personal Data Protection

Already there are important legal safeguards in place in Europe to protect personal data and privacy. The following two requirements must be observed:

1. Limitations on the processing, selling and use of personal data independent of the means which have been used for collecting them. These means generally do not (or did not until recently) involve communications systems and services; classical data collection is by postal mail or commercial desk applications. For the processing, selling and use of these private data, a general Data Protection Directive (95/46) is applicable, and as it is independent of communications aspects, it seems appropriate as it is. It is by and large based on a 'free use unless' paradigm.
2. Limitations on the processing, selling and use of personal data which have been collected or generated through the use of communications services. Since communications systems generate personal data and privacy situations which are not explicitly or knowingly created or desired by their users, a specific telecommunications Personal Data Directive (97/66) complements the general one. However, the legislators designing this Directive had a narrow vision limited to existing switched voice services. Therefore, this legislation needs updating for data generated during communications such as terminal location and e-commerce data, especially in the case of mobile terminals. The current Directive is by and large based on a 'no use unless' paradigm, which is much more restrictive than the general Data Protection Directive.

As of today, data collected for establishing/ensuring communications:

- have to be erased or made anonymous except where they are needed for billing and billing disputes;
- are reuseable by the collecting entity:
    - solely for its communications services, and
    - after the owner's consent.

Data collected at the time of subscription, i.e. not at the occasion of establishing communication, are rather easily reusable for other uses and by other companies, as they fall under the general Data Protection Directive 95/46.

To conclude, convergence and the technical capabilities of IMT-2000/UMTS will lead to a much wider variety of collectible personal data. They range from the precise location of a terminal (i.e. the likely location of a person) to all the use-related personal data one can imagine in mobile commerce transactions. In some countries there exists adequate protection of such data, but in others such protection is lacking or is inadequate. The global nature of IMT-2000/UMTS makes a more comprehensive approach to these problems necessary.

3.4.2.2 Legal Protection of Databases

The value of a database is not merely the sum of the values of the individual data in the database, but also an added value created by the arrangement, organisation and cross-correlation of data in one or several databases. Convergence and the emergence of IMT-2000/UMTS services and mobile e-commerce will raise a series of new issues concerning the legal protection of databases. These issues apply to both wired and wireless networks and need further study by regulators.

3.4.2.3 Intellectual Property Rights

In the future, for example, it will be possible to download music immediately from a IMT-2000/UMTS network at any place and any time. It is questionable whether intellectual property rights can be safeguarded in such an environment. At the international level the World Intellectual Property Organization (WIPO) is in the process of adjusting the international legislative framework to facilitate e-commerce through the extension of the principles of the WIPO Performances and Phonograms Treaty to audiovisual performances. WIPO also works with the adaptation of broadcasters' rights to the digital era and is progressing towards a possible international instrument on the protection of databases. The question remains as to whether the rules will be able to be enforced, and to what extent.

3.4.2.4 Legal Interception and Legal Location

National authorities usually have the right under the terms of regulation or licensing conditions to order public operators to provide legal interception under certain circumstances. Within Europe, a resolution called International User Requirements (IUR) 1995 gives guidelines for this, and similar instruments exist in the United States and elsewhere.

The legal interception schemes should be maintained in principle, but the question of who bears the cost remains. In any case, including these features in the standards will lead manufacturers to implement them in normal commercial versions of their systems, and therefore to reduced costs as compared to mandated a posteriori developments.

Location data must be available to the authorities even when the user has inhibited the location features related to his/her mobile. The principle is the same as for legal interception, and standardised tools will reduce the costs.

3.4.3 HARMFUL AND ILLEGAL CONTENT AND PROTECTION OF MINORS AND PUBLIC ORDER

While public interest objectives relating to the protection of minors and public order have traditionally been recognised in, for instance, the USA and the EU, both at national and

Community level, these have not been recognised in all countries of the world. As the IMT-2000/UMTS network will be global, content which is unsuitable for minors or which disrupts the public order could be available to anyone, any time, any place.

The above also applies to other harmful and illegal content. It is extremely difficult to enforce safeguards in the context of harmful and illegal content on the Internet. The global nature of the platform and the difficulty of exercising control within a given country are leading to solutions which draw on self-regulatory practises by industry rather than on formal regulation, accompanied by technological solutions to ensure that parents take greater responsibility.

3.5 MOBILE INTERNET PROTOCOL

In less than 5 years, in the second half of the 1990s, the Internet grew from 16 million to over 190 million users (see www.nua.com/surveys/how_many_online/index.html). It has spurred billions of dollars of economic growth and enabled the meteoric growth of thousands of companies, from small start-ups to multinational corporations. It has started to transform businesses, governments and other organisations around the world into e-businesses. Yet the Internet revolution is far from complete. Now, you see billions of people using the Internet, whether with a PC, a cell phone, a PDA, or some other type of wireless device, although the dotcom bubble burst only a few years ago.

Mobility, in all its numerous forms, has become the buzzword of our society. An inherent characteristic of wireless systems is their potential for accommodating device roaming and mobility. Everything moves faster and faster. IP seems to be the end-to-end protocol of the future delivery of most services since it exists in the wireline and wireless world, in office extension environments and home networks.

In order to advance the Internet to a new level of efficiency for both networked communications and applications development, a new Internet Protocol version 6 (IPv6) has been specified by the IETF. It fixes a number of problems in IPv4 and, in addition to a number of specialised protocols, offers the following features for cellular networks and users in its 'native' form:

Operators

- auto-configuration
- embedded encryption support and authentication
- embedded mobility
- embedded multicasting
- Internet provider selection
- efficient packet processing in routers
- real-time support
- protocol extensions for proprietary solutions.

End-users

- easy management (auto-configuration)
- efficient address allocation
- improved multicast management
- possibility of renumbering

- efficient network route aggregation
- efficient packet processing in routers
- real-time support.

To avoid the bottleneck of information transfer within the Internet the mobile industry should agree on the deployment of common IP protocols that impact roaming (i.e. the interfaces between core networks) and the communications between terminals and networks in order to achieve maximum interoperability.

Mobile IPv6 (MIPv6) seems to be well suited for cellular networks for a number of reasons. First, the increased addressing space does not require the mobile node to use dynamic address allocation (there is enough addressing space in the foreseeable future for every mobile terminal in the world to use two IP addresses). This could significantly speed up the registration process. Second, MIPv6 does not use the concept of foreign agents, which means that a mobile node must use a co-located care-of address. This address can be acquired by IPv6's stateless address auto-configuration. This method allows a mobile node to add its own link-layer address (which is assumed to be globally unique) to the prefix of the local network to which the mobile node is attached to form the co-located care-of address. In this way, it is not necessary for the mobile node to query e.g. a DHCP server, which also saves time.

Since the mobile node is always using a co-located care-of address, there are no problems with firewalls that implement ingress filtering either. When communicating with a correspondent node, the mobile node's original packet will always be encapsulated into a second IP packet that specifies the mobile node's care-of address as source address. There are no problems with ingress filtering since this is a topologically correct address.

As there is still not sufficient coordination at the international level concerning these matters, the result is that the same problems are being solved in different ways in different countries. In the USA, for example, the High Court of Justice decided that an Internet provider is a post office and that it is therefore not responsible for the content of websites and newsgroups. The court in the UK, however, considers an Internet provider a publisher. This implies that publishers are obliged to take reasonable steps to prevent the publication of offensive material or have such material removed.

3.6 IP MULTIMEDIA SUBSYSTEM: SERVICE ASPECTS

3.6.1 DEFINITION

What is meant by IP multimedia subsystem (IMS)? What benefits does it bring to users, operators and vendors? For a long time, there was no commonly agreed definition of IMS. Although there is still no widely agreed definition of IMS today, the vision of what the 'IMS approach' is designed to achieve remains. However, IMS is now understood to be an evolution of third-generation (IMT-2000) mobile technology that brings:

- the ability to deliver IP-based real-time person-to-person multimedia communication (including voice over IP – VoIP);
- the ability to fully integrate real-time with non-real-time multimedia and person-to-machine communications;

- the ability for different services and applications to interact;
- the ability for the user to very easily set up multiple services in a single session or multiple simultaneous synchronised sessions.

IMS provides a vision of robust, highly valuable services that integrate multimedia activities and allow services to interact with each other, thus enhancing the natural, intuitive process of the end-user, in whatever network he/she operates, whether it is fixed, mobile or roaming. The resultant integration and interaction of media types opens up new possibilities for far richer services than are available today.

The power of this interaction and integration is significant. It is the integration and inter-action of services that can turn a frustrated user attempting to manage multiple activities on a small device into a satisfied customer using mobile services involving multimedia activities in a seamless fashion, allowing a natural, intuitive process. This is what end-users expect from mobile data services, and they are willing to pay for it.

Still, qualitative differences in user experiences are difficult to quantify. Since IMS deployment is an option for operators, the need for IMS is a subject of debate among industry players, especially since many services can be emulated to some extent without IMS. How-ever, consumers have clearly demonstrated a willingness to pay for integration of services. In the mobile environment, ease-of-use has always been highly valued, and users will expect mobile data services that offer comparable convenience. To obtain high adoption levels, the mobile industry must also make mobile data services natural, intuitive, convenient and easy to use.

For the mobile operator, IMS provides the potential for interoperability of mobile and fixed networks and a robust creation platform, which in turn can be used to increase competitive advantage. IMS provides a standardised solution for enhancing the end-user experience that cannot be easily duplicated by any known technology today. In addition, a vision of IMS is portrayed that includes interoperability between fixed and mobile IP networks that will further improve the possibility of a common satisfactory end-user experience across services, networks and devices.

3.6.2 THE IMS VISION

IMS is designed to integrate mobile communications and Internet technologies, bringing the power and wealth of Internet devices to the mobile environment. IMS enables interoperability between fixed and mobile networks and so holds the promise of seamless converged services.

Two features of IMS are of fundamental importance:

- IP-based transport for both real-time and non-real-time services;
- introduction of a multimedia call model.

Both standardisation groups 3GPP and 3GPP2 have selected the IETF Session Initiation Protocol (SIP) for multimedia call control.

Regarding the two features named above, this combination is not only extremely powerful, but also radically changes the communications process and could introduce an entirely new dimension to telephony.

Until now, communication services in telephony such as voice calls or data transfers from a server have essentially been restricted to one service per bearer. More complex communications processes can be constructed and deployed, but these require multiple bearers, leading to an inefficient use of network resources and an unwieldy user interface. IMS removes these constraints.

IMS enables communication sessions to be established between multiple users and devices. It allows multiple services to be carried on a single bearer channel. IMS allows the integration of real-time and non-real-time services within a single session and provides the capability for services to interact with each other.

IMS gives users the ability to set up separate services within a single session, multiple simultaneous synchronised sessions, or multiple simultaneous single sessions of unrelated services. It will provide a significantly enhanced user experience and promises to enable radically new services. IMS thus provides three key features to end-users:

- integration of services
- interaction of services
- presence.

Integration of services is the ability to dynamically modify the media types active during a multimedia session call. The range of media types in use is dictated only by the media capabilities of the user terminal. In this way, IMS integrates into a single session what are today separate services. For the user, this single session capability means users can multitask. This means that they do not have to terminate a voice call (or place it on hold) in order to send a text message or a video clip.

The third feature, 'presence', is the ability to know if a user is currently online. Though seemingly a simple capability, in an IMS environment, presence and its extensions provide a set of extremely powerful service enrichment capabilities. For example, presence would permit a user to launch a 'when available' conference call with multiple users and have that call established when all required parties are available. The incorporation of location-based information with presence in IMS-based networks allows innovative services, which leverage the proximity of a user to specific locations or users, and can be available even when roaming.

3.6.3 INTERNETWORK INTEROPERABILITY

The ability to deliver the full promise of IMS requires the solution of some outstanding technical issues. The provision of appropriate quality of service for real-time services in a packet-switched environment is one. Inter-working between E.164 numbering schemes and IP addresses is another.

Within the IMS, services and networks are completely distinct. Service creation is therefore accessible by a broad community of developers – a similar situation to the development of Web content and access in the fixed Internet environment. IMS provides the tools to deliver what the PSTN intelligent network concept only promised.

A key feature of IMS is that the control and intelligence is largely transferred to the edge of the network, or even into the terminal device itself. Such an architecture facilitates the provision of connectivity between networks. In principal, IMS enables interoperability between 3G mobile networks and fixed networks such as the PSTN and the Internet.

In particular, IMS moves real-time services, such as voice telephony, from the circuit switched (CS) domain to the packet switched (PS) domain, supported by Internet protocols (IP) rather than the GSM MAP protocols which have their origins in ISDN. It means that all services will eventually be delivered via one integrated network rather than two overlaid networks, and this is also expected to bring cost savings in equipment, customer care and network management.

At first sight, IMS appears to be only relevant to the core network. Indeed, in 3GPP, IMS started off as a proposal for a new all-IP network architecture concept and it was only later that requirements for services were considered. However, the 3GPP documents show that it is much more of an end-to-end system understanding: the 3GPP Service Requirements Specification states that IMS is a complete solution for the support of multimedia applications (including voice), and that the solution includes terminals, radio access networks (RAN) and the evolved packet core network. What IMS is *not* essential for, is ordinary access to the Internet, Web browsing, email, messaging and text, and even advanced services like streaming video and audio. All these are server-to-person services rather than person-to-person or non-real-time services.

3.6.4 VOICE OVER IP

VoIP is more than just IP voice. VoIP brings the ability to deliver IP-based real-time person-to-person multimedia communication. Real time means that there is no delay between what is sent and what is received. Multimedia means video telephony, audio or speech telephony, graphics and text. Most importantly, this can be done simultaneously in one session. It is not just convergence in the old sense, where fixed and mobile networks are connected to each other in the sense that it makes no difference from where you start your call and where it may end: fixed or mobile.

Real-time conversation can be combined with non-real-time information, and this can be a multiple simultaneous sessions, or – unlike GPRS or ordinary 3G – a single session with multiple media. The 3G packet capability or even GPRS are sufficient for these services. However, it is important to realise that these conventional Internet applications can be implemented by means of IMS, and it may be cheaper and simpler to implement them this way. Conventional Internet applications can also be combined with IMS to further increase their potential.

As an example of how real-time telecommunications services can be integrated with information services, consider directories, probably the most commonly used feature on a phone after speech. In today's mobile phones, directories are offline, held in the SIM card or stored in the terminal. Unfortunately, quite often when using the directory, you find that the information quickly gets out of date and the attempted call fails. This is because area codes and people's numbers seem to change frequently. With IMS, however, directories can be online, for example stored on a server in the network. The online directory is permanently accessible, because the 3G packet data service is always on. It will always be up to date because it will know if a subscriber's area code has changed, for example. This is because the online directory will be updated from information in the various databases in the network – wherever they are.

Of course, the internal directory in the phone can also be synchronised automatically by the online information. The IP multimedia call processing (SIP) can interact directly with the directory server or any other server on the network to enable it to set up a call or sequence of calls. Therefore, e.g. IMS will integrate the telephony services (the call set-up process) and the information services (the directory).

An excellent example is the presence service applied to telephony. Here, the call processing interacts with a presence server. It can let you know whether the subscriber you want is available (Figure 3.5).

In the example in Figure 3.6, on the screen, you see that the called person is unavailable by phone or may not want to be disturbed – for example by a call waiting signal, but that he is

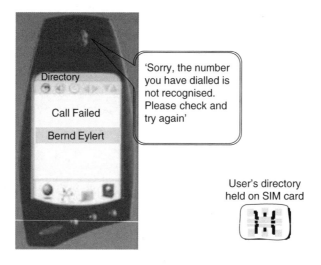

Figure 3.5   Presence service applied to telephony. *Source*: UMTS Forum (2002c).

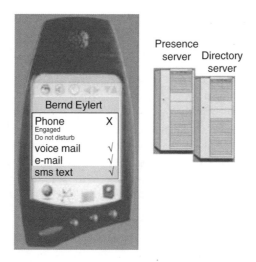

Figure 3.6   IMS integrated services. *Source*: UMTS Forum (2002c).

available on voice mail, email or text messaging. The SIP call processing can interact with the directory or with any other application server. IMS would integrate telephony services and information services.

Figure 3.7 is another example of advanced interactive information services that can be integrated with telephony services. You can use advanced online search engines to find subscribers – just type in the name of the person you want to contact, together with some details, just like a Web search engine. The SIP call processing can interrogate the directories directly and set up the call automatically. The user never needs to know a phone number again. You could do such a search on an online search engine using conventional GSM and GPRS or 3G, but without the integration provided by IMS the result would only be the display of a phone number. The user would have to make a note of it and re-enter it on the keypad. So with IMS, everything is automated and easy to use, bringing a much better user experience. It might be possible to create such a service without IMS, but it would need terminal specific software, leading to major compatibility problems. A good example of that phenomenon was multimedia messaging in the early days of MMS, which in the beginning was not sufficiently standardised, even not harmonised between manufacturers' equipment (inter-terminal interoperability).

This principle could be extended to public directories like Yellow Pages. SIP call processing would enable two users to have a discussion (speech) while simultaneously viewing a Yellow Pages entry, with one user controlling the information displayed on the other user's screen as well as on his own. This is an example of multiple synchronised sessions enabled by SIP and IMS.

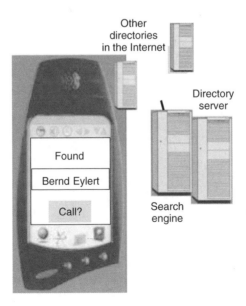

Figure 3.7 Advanced interactive information service. *Source*: UMTS Forum (2002c).

With IMS and a videophone you could send a video picture to a cable TV – for example, a school sports event. The rest of the family could then watch it on the large screen. You could also remotely switch on your home video recorder from your phone. This can be offered by IMS in future because everything listed below will be based on IP transport, signalling and control (Figure 3.8):

- the mobile terminal and IMS network;
- the VoIP-enabled telephone network;
- the digital cable TV network;
- the digital set-top box, now IP multimedia enabled;
- the Internet itself;
- WLAN and Bluetooth;
- IP-addressable domestic appliances.

IP is being adopted as the universal 'language'. Its capability will be available on all devices. Therefore, it is inevitable that the mobile networks will move to IP and IMS.

3.6.5 IMS IN SUMMARY AND CONCLUSION

An IMS-based network is superior to a non-IMS-based network in many areas. The key advantage of IMS is that it incorporates all the capabilities described earlier into a standards-based implementation, thus enabling interoperability between IMS-based networks and fixed

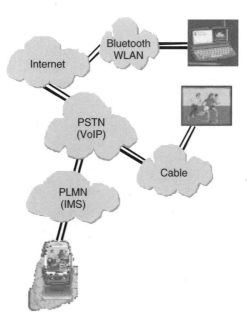

Figure 3.8   The comfortable network. © UMTS Forum (2002c).

IP networks. One could argue that a mobile operator could, at a considerable expense, create a non-IMS-based network that provides IMS-like features at a point in time. The technical feasibility exists. However, new end-user desirable services created for this propriety network would not likely be fully interoperable with other networks. This would limit users' ability to fully utilise the new IMS-like services as they roam from one network to another.

Additionally, third-party service developers in a non-IMS environment would be less aggressive in their development of new services, because the market potential has been limited to the user base of propriety network mobile operator. Also, the cost and challenges of supporting multiple propriety networks are prohibitive. Therefore, the business and market feasibility of the propriety non-IMS-based network and its new service capabilities are greatly reduced.

Another significant benefit of IMS is the establishment of a platform environment to rapidly create, deploy and modify new advanced mobile services for end-users. The value of IMS that integrates voice with data IP services comes from integration – not at the network layer, but at the services layer. In so doing, the rapid new services deployment and customisation capabilities of the fixed Internet, which have brought innovative multimedia information services to end-users, can be brought to the mobile world.

3.7 NAMING, ADDRESSING AND OTHER IDENTIFIERS

Naming, addressing and identification is a very specialised subject, and some of the basic principles and requirements are now considered. The difficulties in organising the customer-oriented items in relation to the services and applications are explained, the procedures, how customers can reach one another, the services and applications they have subscribed and the billing process they will be charged for. This section is based on the UMTS Forum's report no. 12 (UMTS Forum, 2001c). In order to make reading easier all technical-related aspects on this subject are moved to Chapter 4.

3.7.1 NAMING

3.7.1.1 Introduction

A name is a 'combination of characters and is used to identify end users. (Characters may include numbers, letters and symbols)' (ITU-T Recommendation E.191).

An end-user is a 'Logical concept which may refer to a person, a persona (e.g. work, home), a piece of equipment (e.g. phone), an interface, a service (e.g. freephone), an application (e.g. video on demand), or a location'.

A name is distinct in function from an address. Addresses are essential for communication as the end-points always have to be identified in a way that can be used for routing, but names are not essential. Names are added for some services to make it easier for users to identify the distant end-point or to provide an identification system that is independent of the structure of the networks or the current location of the entity to be communicated with.

There are two common naming schemes:

- E.164 names (numerical strings), defined by ITU-T Recommendation E.164[3] – The International Public Telecommunication Numbering Plan.
- Names of the form 'user@domain' defined by RFC 1035 – Domain Names – Implementation and Specification.

The choice of naming scheme is related to the nature of the service because a service description needs to specify which type of name is used. This is important, because:

- users need to know how to identify their correspondents;
- the choice of naming system determines the set of potential correspondents that can be reached;
- interconnected networks need to have a common method of identifying communicating users.

For many services names are used as the identification system, but some services allow addresses to be used as an alternative to names (e.g. http allows users to identify Web sites by IP addresses or domain names) and some services use addresses only.

In the past, services and hence name types were related to technology. For example, telephony could be provided only on circuit switched technology and Telex had its own naming scheme and its own technology. However, third-generation mobile technology is designed to support multiple services and hence, there is the possibility of supporting more than one type of name.

3.7.1.2 General Naming Issues

Relationship of Names to Services and Users

Normally each service specifies a single type of name that is used. However, a type of name may be used by several different services. Sometimes these services are distinguished by different ranges, as is the case of E.164. Where there is a separate method to identify the service separately, the same value of a name of the same type may be used for different services. For example, a given E.164 number may be shared for both telephony and fax if the terminal can distinguish the services, or the same value of user at host may be used for both email and SIP-based services.

A user may therefore have several names for different services, e.g. GSM users have three different E.164 names for voice, fax and data services on GSM. This is not very user friendly because business cards become cluttered up with the different names. The long-term objective is to work towards having one name per person for private use and perhaps a separate name for business use. However, this goal is constrained by the need for compatibility with existing systems.

A single name can also support several different users. Examples are an E.164 name for a telephone service in a house shared by several occupants or an E.164 name used by a call centre or an email name used by several people who fulfil the same function in an organisation (e.g. sales@company.com).

---

[3] ITU-T Recommendation E.164's numbering plan is a mixture of names and addresses. It started primarily as an addressing system, but has migrated to become more of a naming system, because location and operator portability are functions of names rather than addresses.

Although one name of the form 'user@domain' may be used for several services, a given name may be used only for services supported on a single service provider because the name will be mapped by a domain naming scheme (DNS) to an IP address of an interface operated by the service provider. Therefore a user who wishes to use one service provider for one set of services and another for another set will need different names.

Information in Names

Although the main function of a name in telecommunications is to provide unique identification of the end-user, considerable additional information may be provided. This additional information may be more or less explicit and may not always be beneficial. Examples are:

- the type of service (e.g. a number range in E.164 dedicated to mobile; this may also indicate the expected tariff level);
- the likely tariff level (e.g. freephone);
- the location (e.g. country code or area code);
- international networks (e.g. shared CC 882 in ITU-T Recommendation E.164);
- the service provider (e.g. the value of 'domain' as in john_smith@compuserve.com).

A further form of information is an association from digits to the alphabetical characters shown on the keypad of a telephone (defined in ITU-T Recommendation E.161). This form of dialling is used commonly in the USA to support advertising, e.g. 'dial 0800hotpizza', but has become more common in other areas of the world as well (e.g. Europe and Asia).

The provision of information limits portability. For example, location portability is restricted to the area specified by the location information in a name, although there are exceptions such as the use of country identifiers by organisations operating outside the country concerned. Several countries assign Internet names to organisations outside their boundaries (e.g. umts-forum.org or itu.int).

In E.164, the identity of the number block used does normally indicate the service provider but this information is not recognised by most users and so service provider portability can be supported.

The provision of explicit service provider information that is perceived by the user, such as when the value of 'domain' includes the identity of the service provider, is not readily compatible with portability between service providers. Names of the form 'user@domain' can be portable between service providers without being misleading only if the value of 'domain' is not the name of the services provider.

Most companies have their own domain name and the domain name can be hosted either on their own server or on that of an ISP. In contrast, a name for an individual such as john_smith@compuserve.com links the user strongly to the service provider Compuserve. Some countries have started to provide special domains for individual users who want portability but do not want to obtain their own domain name.

Service Types Supported on IP

Services supported on IP can be classified roughly as:

- any-to-any
- client–host.

An any-to-any service provides communication between end-users. Examples are public telephony, facsimile and email. These services may use client–host type relationships in their provisioning (e.g. access to an email server), but the main focus for the user is on communications to another end-user, therefore the naming system needs to identify the end-user uniquely. Any-to-any services typically use E.164 names or Internet names of the form user@domain.

Client–host services focus on access to facilities provided by a host, such as access to information on a Web page. Client–host services make less use of E.164. Some people argue that dialling into an information line with voice or tone response is a client–host service as well.

Backward Compatibility

As naming determines the set of subscribers that can use a service, and the value of any-to-any telecommunication is a very strong function of the number of users, backward compatibility is extremely important. It would be very difficult commercially to launch a new service that is an alternative to an existing service with a different naming scheme as users would wish to be able to reach all the users of the other service. New naming systems can, however, be introduced for new services, and this has enabled the name type 'user@domain' to be introduced for services such as email.

Human User Aspects of Names

Where communications are established by humans rather than machines, the human-related aspects of names have some importance. This aspect is enhanced in directory services (e.g. see Section 3.6).

Communications (not just telephone but also data, fax, email etc.) can be categorised into three types:

1. regular and frequent calls between informal or formal groups (colleagues, family, other activities), for which the main issue is easy call establishment;
2. occasional calls to advertisers or major organisations or services (transport, government, large retailers), where the names need to be easily memorable;
3. occasional and largely random communications to other destinations.

Type 1 requires the storage of names under customised strings as in personal address books. This requirement can be handled by developments in terminals. It does not require standardisation.

Type 2 requires human memorability. In E.164 this involves golden numbers or the relationship of a number to a memorable alphabetical string. However, the scope for using these associations in E.164 is severely limited because of the problem of duplication, especially for individual names. This is not a technical limitation but a limitation intrinsic in the established use of names.

Type 3 requires good directory services or search engines so that terminals can find the correct number quickly from incomplete searching information when a call is initiated. Terminals should be capable of initiating connections by using the numbers or names obtained without the user having to re-enter them.

Issues for UMTS

Internal naming of ISPs: Access Point Names

For both GPRS and UMTS the intention is to use an internal naming system for identification of the ISP that a mobile wishes to log-on to. These names are called access point names (APNs). More information is given in Section 4.5.3.

This naming system mostly uses names that are already registered for public use but the GSM Association registers names in other cases. The GSM Association also assembles and provides the information for DNS servers to be used by the operators.

Presumably the GSM Association will continue this role for UMTS.

External naming

External naming concerns the choice and allocation of names to UMTS users so that:

- other UMTS users and other non-UMTS users of the same services can communicate with them;
- UMTS users can present appropriate calling identification in their communications.

External naming is primarily an issue for the ISPs on the edges of the UMTS networks. These ISPs will serve the UMTS customers and register their name–IP address pairs in the global public DNS if the UMTS users are to be publicly accessible. Some UMTS users will choose to operate in private or restricted virtual networks and so will not be reachable from the external world.

The main issue for external naming is the choice of naming system. This is a separate issue for each service. Internet names are likely to be preferred for established Internet services and other new services. For services with a large number of users on switched circuit technology, E.164 is likely to be preferred.

E.164 issues

E.164 is the numbering scheme of the ITU-T Recommendation on Telephone Network and ISDN Operation. E.164 defines the structure and maximum length of its numbers and lists the values of the first 1, 2 or 3 digits that have been allocated to particular countries, global services or networks.

The top part of all E.164 numbers (country code, global service code or shared network code) are allocated by ITU-T Study Group 2. The remaining part is allocated by the next authority level, e.g. the national numbering authority for country codes. Most numbers are allocated in blocks (typically 10 000 numbers) to network operators or service providers, which then allocate individual numbers to customers. In some cases numbers are allocated directly by the numbering authority or its agent to the end customer. True E.164 numbers are always capable of expansion into the full international form and should be reachable from any country.

E.164 names are used by the following services:

- telephony
- fax
- circuit switched data
- mobile short message service.

Telephony and fax are not normally segregated into different number ranges. In contrast, data and SMS numbers may be segregated at a national level but this is not recognised by callers.

In addition, within the general heading of telephony, E.164 names are used for:

- geographic services (the most common use)
- mobile services
- satellite services
- paging
- global freephone
- global shared cost services
- global premium rate services
- personal numbering services including Universal Personal Telecommunications (whose numbering is defined in Recommendation E.168) with local, regional or national options.

E.164 numbers are be used for both names and addresses, although with the growth in demand for operator portability they are increasingly used as names rather than addresses, i.e. they are not related exclusively to a particular network.

Although all public telephone numbers are coordinated with E.164 so that there can be no dialling ambiguity, not all telephone numbers are formally recognised as E.164 numbers. The following are examples of numbers that are not formally part of E.164 but fit into the national dialling plans alongside true E.164 numbers:

- short codes (e.g. 100 for the operator)
- national freephone or shared cost numbers not reachable from outside the country concerned
- routing numbers that cannot be activated by user dialling (e.g. location routing numbers in the North American number portability solution).

For public voice telephony, E.164 is the obvious choice because of the commercial imperative for compatibility with PSTN/ISDN users. In many countries there are also regulatory requirements for the presentation of calling line identity (CLI) at least to interconnected operators for the support of access to emergency services and malicious call detection, and the systems used for these purposes are likely to remain limited to handling E.164 names for the foreseeable future.

Within E.164, the three main issues are:

- Which number range should be used for UMTS?
- Should portability be required between GSM and UMTS?
- How should new UMTS numbers be allocated – in blocks through operators and service providers or directly to users (individual allocation)?

Choice of number range in E.164

If portability is required, then range and number portability are related to the fact that UMTS will have to use the same number range as GSM. Furthermore, the choice of technology for implementing portability may facilitate call routing where there is individual allocation.

Two forms of numbering and charging are used for mobile services:

- Calling party pays services with special mobile number ranges, as used very widely in Europe.
- Called party pays for the mobile termination with number blocks from the scheme for geographic services, as used extensively in the USA.

For 'calling party pays', there are three strong arguments for allocating numbers for telephony on UMTS from the same number ranges as GSM, provided that the services and tariffs are similar:

- The service (mobile telephony) is essentially the same and numbering should relate to services rather than technology.
- Those operators with both GSM and UMTS licences are likely to want to be able to provide portability for existing customers who wish to migrate from one technology to another. Therefore UMTS must be allowed to use the same number range.
- New market entrants, who have only a UMTS licence, need number portability in order to be able to compete for customers of the incumbent network operators.

Where new services or tariffs are introduced that have no equivalent to e.g. GSM, the use of different number ranges could be advantageous or essential. Where multimedia services include voice service that is compatible with telephony, the choice of number range will need careful consideration as some customers may wish to upgrade from conventional telephony service to multimedia while keeping the same number.

Number portability
The European Commission has urged all Member States to make mobile number portability happen. In most Member States it has been introduced, although it took some time. The introduction of UMTS including number portability gives new entrants who do not have a GSM system the opportunity to compete more effectively for GSM customers. Number portability in GSM has included both portability between operators and portability between those service providers which resell airtime, where number allocation involves services providers.

Number allocation
Numbers for subscribers are normally allocated in blocks to operators and, where appropriate, sub-allocated in blocks to service providers. Individual number allocation has been introduced in a few countries for services such as freephone services, shared costs, premium rate and personal numbering services. Regulators may wish to consider whether there should be any requirements for individual number allocation for mobile services. Individual allocation has most value where users may wish to have memorable or branded numbers. It is unlikely that there will be a very strong case for individual number allocation for UMTS.

For 'called party pays', the numbering arrangements are part of the geographic scheme and should conform to all the requirements on those schemes, including portability with fixed services.

Numbering allocation for data-only terminals may require E.164 numbers (Mobile Subscriber Integrated Services Digital Network, MSISDN), although they do not require any incoming call. This requirement could increase significantly and unnecessarily the demand for E.164

numbers for mobiles. A possible solution could be the use of a separate private numbering scheme, if access from the public networks is not required. More studies are needed on this issue.

Personal numbering

Personal numbering enables calls to be delivered to any terminal on any network according to information supplied to the personal numbering services by the called user. Therefore users who are called with a personal number may have calls delivered to their mobile extension. This possibility will continue for 3G/UMTS systems. Universal Personal Telecommunications (UPT), whose numbering scheme is defined in ITU-T Recommendation E.168, is highly standardised.

Personal numbering provides an overlay of personal numbers on top of existing network-specific numbers. It does not affect the numbering arrangements for the networks on which it is provided as it does not replace them.

Mobile Station Roaming Numbers

Mobile Station Roaming Numbers (MSRNs) come from the E.164 numbering space although they are not public numbers (they are coordinated E.164 numbers). MSRNs are used by GSM operators for routing incoming calls to mobiles. The MSRN is allocated temporarily to a visiting mobile by the visiting MSC (VMSC). This arrangement will continue for the support of circuit switched services (3GPP Release[4] 5). The introduction of UMTS with new operators entering the market will increase the demand for MSRNs but national numbering authorities should be able to satisfy the additional demand without undue difficulty.

Internet name issues

Internet names are the names used by the Internet community. This naming system is administered by the Internet Corporation for Assigned Names and Numbers (ICANN). The names are written normally as 'user@domain' where 'domain' is the domain name. The domain name is composed of a series of strings separated by '.'s, e.g. 'companyname.co.uk'. The most significant component is to the right and is called a top level domain (TLD).

TLDs are structured into two types:

- organisationally structured
- geographically structured.

Organisationally structured TLDs are called generic TLDs (gTLD). Domains of a similar type are grouped under one TLD, e.g.:

- .com for commercial sites
- .org for non-commercial organisations
- .gov for the US government
- .net for network-related groups
- .edu for educational institutions
- .mil for the US military
- .int for international organisations
- .arpa for Internet infrastructure.

---

[4] 3GPP calls a dedicated development phase a 'release' to structure the standardisation process.

Geographically structured TLDs are called Country Code TLDs (ccTLD). In this category, sites are grouped by country, e.g.:

- .ca for Canada
- .de for Germany
- .fr for France
- .it for Italy
- .jp for Japan
- .uk for the United Kingdom.

ICANN is responsible for whether, how and when new TLDs are added to the existing ones. In October 2000, ICANN approved seven new gTLD of the following types:

- .aero for the air transportation industry
- .biz for commercial businesses
- .coop for cooperations
- .info
- .name for individuals
- .museum for museums
- .pro for professionals.

TLDs may be restricted to a specific use, charter (e.g. .gov for the US government, .museum for museums), or open to any use (e.g. .info). Some of the former restricted TLDs (e.g. .com, .org) are now handled more openly.

3.7.2 ADDRESSING

3.7.2.1 Introduction

An address is defined as a 'string or combination of digits and symbols that identifies the specific termination points of a connection and is used for routing' (ITU-T Recommendation E.191). An address identifies the interface at which the connection is to be delivered without regard to whether the connection continues beyond that interface. Addresses contain location information and in telecommunications this is expressed in terms of the network structure in order to achieve as high as possible a degree of aggregation, which reduces the complexity of routing tables in switches or routers.

Addresses differ from names where addresses contain explicit network information and this information is what makes them usable for routing. In order to route a call or a packet, the called name must be translated into an address which identifies the location and so can be used in the routing process. The extent of the information to support routing will vary with the type of address. Routing information may need to be supplemented. When a name is ported from one location or one service provider to another, the address associated with the name changes.

Unfortunately, the distinction between name and address is not followed consistently and entities that are names, or closer to names than addresses, are often spoken of as addresses. Examples of this incorrect usage of terms include 'email address' and 'SIP address' where the 'domain' may not denote a service provider or network operator but be the user's own domain. A uniform resource locator (URL) pointing to a company's Web page is often called an Internet address, but it is actually based on a domain name.

Three types of addresses need to be considered for UMTS:

• IP addresses
• mobile station roaming numbers
• routing prefixes for E.164 numbers.

3.7.2.2 IP Addresses

Use in UMTS
IP unicast addresses identify interfaces, which are end-points for IP packets (multicast addresses identify groups of unicast addresses). IP addresses are used in three different ways in UMTS. They are used for:

• end-points within the GPRS/UMTS network infrastructure (e.g. SGSNs and GGSNs), where the addresses are assigned and managed by the GPRS/UMTS operators;
• mobile terminals connected across the mobile network to an ISP, where the addresses are assigned and managed by the ISPs;
• mobile terminals in multimedia services (Release 5 only), where the addresses are assigned and managed by the CSCF, although details are not yet specified.

The use in the network infrastructure is not visible to the external Internet world in GPRS and UMTS Releases 3 and 4; visibility may be possible in Release 5. These addresses are used by the internal GPRS tunnelling protocol for setting up tunnels between the SGSN and the border gateway. Each tunnel supports communications between a specific mobile and the ISP that it is logged-on to.

Assignment of addresses to mobile terminals is necessary to enable them to communicate with the external Internet world. In Release 5 the IP addresses are assigned to mobile terminals by the GGSN that they are using. Communications to and from the mobile use this address, which is carried without alteration or inspection through the tunnel.

IP addresses are divided, in principle, into two parts:

• the identity of the network (e.g. the network part);
• the identity of the interface attached to the network (the host address, which is the destination of the IP packet).

The range of addresses allocated to ISPs may be chosen to provide aggregation, i.e. ISPs that are connected to the same transit (backbone) operator may have adjacent allocations.

The identity of the interface is assigned and managed by the network operator.

The Power of IP Addresses
There are two versions of IP protocols, whose address formats differ significantly:

• IPv4, a 32-bit address
• IPv6, a 128-bit address.

IPv4 has a 32-bit address size which is used throughout the Internet, but which is considered to be in short supply. Although initially IPv4 was considered robust and scalable (it had the ability to uniquely identify over four billion nodes), the rapid increase in demand coupled with the inflexibility of assigning addresses in strict 'Classes' led to problems. While the introduction of classless inter-domain routing (CIDR) enabled these restrictions to be overcome, questions are now being asked about the ability of IPv4 to accommodate future demand.

IPv6 is a 128-bit address, which is just starting to be used and should provide more than adequate capacity for the future. Compared to IPv4, IPv6 allows 10^{38} possible addresses, which is an almost infinite number.

When eventually IPv4 reaches exhaustion, new allocations will be possible only from IPv6. This will mean that equipment with only IPv6 addresses will be able to communication only with other equipment that has IPv6. Communications with the IPv4 world will not be possible. This will be a significant commercial issue, therefore the introduction of IPv6 should be encouraged in order to become as widespread as possible before IPv4 is exhausted so that the loss of compatibility will be minimised.

With the introduction of IPv6 the IPv4 address size will be increased to 128 bits to allow a transition from IPv4 to IPv6. Currently the amount of Class A address space (classes are described in Section 4.5.4) remaining is estimated to be in the order of 1.25×10^9. Of the 128 Class As, 61% are not yet allocated. A Class A, being the largest IPv4 address block that can be allocated, represents 16×10^6 IPv4 addresses. This huge remaining address range is thus available for allocation. There are also many large address blocks assigned to organisations that are still not used (i.e. IPv4 addresses that are not actually advertised on the Internet), or that are only partially used. It should be acknowledged that some address space will remain unusable due to built in network redundancy and the immense difficulties that would need to be faced through network reconfiguration. Estimates of the remaining useful life of IPv4 range from 5 to over 20 years.

Choice of IP Version

The choice of IP version is an important issue for operators. The GPRS infrastructure and terminals will be based on IPv4 as IPv6 was not ready in time.

Operators are therefore faced with decisions about the choice of version for their core network. The main decision is when to introduce IPv6 into the core network and when to start parallel operation of IPv4 and IPv6.

From the perspective of the UMTS operators and their core network, they will need to start to introduce IPv6 when they start to implement IP multimedia services under Release 5. At this point they may wish to start introducing dual stack capabilities for the other core network functions. Unless operators perceive any special advantages in IPv6 for the core network, they may choose not to introduce it before they start to support IP multimedia.

A related issue for operators is whether there needs to be any coordination between different operators over the introduction of IPv6 for the core network. Such coordination does not appear to be necessary as implementation of IPv4 will remain mandatory and therefore operators will always have the option of using IPv4 for tunnels from an SGSN in one network to a GGSN in another.

Temporary or Permanent Assignment

Allocations to terminals may be either permanent or temporary. Temporary allocations are normally made by ISPs to terminals on dial-up fixed network access because temporary allocation allows addresses to be reused and reduces the number of different addresses that an ISP needs. Some applications may need permanently allocated addresses to facilitate security and other functions.

GPRS and UMTS will provide an 'always on' capability and terminals will therefore use IP addresses for a significantly larger proportion of time on average, and this will reduce the reuse factor to a small figure, with the result that reuse may no longer be a reason for temporary assignment.

Aggregation of IP Addresses

Aggregation is a very important issue for IP addresses. Aggregation is the practice of allocating adjacent blocks of IP addresses to networks that are connected to the same point in a backbone network. The objective is to reduce the size of routing tables in routers. For example, if allocations are aggregated then all routers other than the ones connected directly to the networks concerned can store a single address range for several destination networks. If there is no aggregation, they need to store separate ranges for each network. Mobile terminals may set up a packet data protocol (PDP) context with either their home ISP or any ISP on a visiting mobile network. Allocations to core network equipment will almost certainly be permanent.

A permanent assignment of an IPv4 address to a mobile would not work in this case because packets to that address would have to be routed differently depending on which ISP the PDP context is with, and this would defeat aggregation and complicate routing. Therefore for IPv4, it appears that temporary assignment will be needed.

The situation for IPv6 may differ to some extent because the address length is sufficient to support both a globally unique permanent terminal identifier and separate temporary routing information. This possibility is under study in IETF and a much higher proportion of IPv6 addresses may be permanently assigned.

3.7.3 OTHER IDENTIFIERS

3.7.3.1 International Mobile Station Identifier and Mobile Network Codes

The International Mobile Station Identifier (IMSI) is defined by the ITU-T E.212 Standard and is used as the primary identification of the SIM card. The IMSI is used for network internal purposes, including the logging-on procedure and billing, and is not normally visible to the subscriber. IMSIs are network-specific, because they contain the Mobile Network Code (MNC). The structure of the IMSI is shown in Figure 3.9.

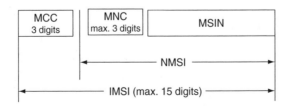

Figure 3.9   Structure of the IMSI showing the MNC. © UMTS Forum.

Ideally, an MNC identifies one specific mobile network (with the introduction of number portability this clear distinction has been lost, because customers now can carry their number to any operator). MNCs are used in both second- and third-generation mobile networks for identification and routing purposes. These purposes include:

- identification of the home network of a roaming mobile during registration;
- preparation of billing information relating to roaming mobiles;
- Analysis in support of fraud detection and prevention.

It is probable that other future mobile or converged (fixed–mobile) services will require these E.212 resources as well.

The E.212 standard gives the option of using either 2-digit long MNCs together with 10-digit long MSINs or 3-digit long MNC together with 9-digit long MSIN. The reason for this is coordination of the European practice with only 2-digit MNC and the US practice with 3-digit MNC.

Whereas MCCs are defined by ITU, the administration of MNCs is up to the National Regulatory Authorities (NRAs). It is also up to the NRAs to grant 2- or 3-digit long MNCs. Therefore the number of MNCs allocated and the number of digits used vary per country.

Demand for MNCs
MNCs are needed by:

- each UMTS network operator with radio licence;
- each UMTS virtual network operator who operates an HLR, but does not operate a radio network.

Because of spectrum considerations, the number of radio licences to be issued will be low in countries with national services but could be larger where local-only operators are licensed. Therefore the main need for a substantial number of MNCs will come in countries where virtual network operators are licensed. Countries that have national systems and do not license virtual operators are unlikely to experience a shortage of 2-digit codes. However, personal numbering and mobility for customers of fixed networks may lead to the use of MNC codes by operators of fixed or technology-independent services and these developments may create pressure for an earlier move to 3 digits than would be necessary otherwise.

Some GSM operators that will also become UMTS operators may wish to use the same MNC value for both technologies since, for example, they may wish to enable their subscribers to use dual-mode terminals with a single SIM. These operators may provide UMTS coverage in only the denser parts of their GSM coverage.

It is not expected that service providers would need MNC codes. Service providers normally obtain blocks of IMSIs from operators for allocation to customers.

IMSI Allocation
The current system of IMSI allocation is inefficient and may lead to premature exhaustion of existing MNCs. Currently blocks of IMSIs are allocated to service providers by operators through the distribution of SIMs. Individual values are sub-allocated when SIMs are given to a customer. There is a high degree of churn in the mobile market and many customers change

operator every 1–2 years. This churn is likely to increase with mobile number portability being introduced or improved in many countries. When a number is ported, although the E.164 number does not change, the IMSI changes. A customer may have already left a trail of several relinquished IMSIs. Thus, churn will increase the rate at which IMSIs are consumed and the space within MNCs is used. At some stage a mechanism for recovering and reusing IMSIs after an appropriate sterilisation period may be needed.

3.7.3.2 International Mobile Equipment Identifier

The International Mobile Equipment Identifier (IMEI) identifies the mobile terminal. It is used for tracking stolen terminals and for fraud prevention. It has also been used in relation to type approval. The GSM Association has developed assignment rules for the IMEIs and these identifiers are registered by the Association. The current assignment rules and principles are under review, in view of the implementation of the RTTE Directive, which changes the type approval arrangements. The use of the IMEI may also be affected by the voluntary certification scheme.

The current IMEI format is structured in the following way:

- Type Approval Code (TAC): 6 digits. The first 2 digits constitute the code allocated to Notified Body = Reporting Body Identifier (1900 MHz phones in the USA and test terminals have different coding).
- Final Assembly Code (FAC): 2 digits.
- Serial Number: 6 digits.
- Check digit.

These digits are presented in binary coded decimal (BCD) format.

The current IMEI allocation is based on the involvement of notified bodies. IMEI has been so far a regulatory requirement. However, the RTTE Directive will remove this regulatory requirement. The industry considers IMEI an important market requirement, so IMEI must be secured. Therefore it is envisaged that the allocation should be transferred to manufacturing industry. In the future it will contain information on the manufacturer as well.

The review of the IMEI system should enable:

- a fair, transparent and secure allocation procedure, meeting the interests of network operators and manufacturers;
- a common scheme to be used for GSM and UMTS terminals as multi-mode terminals are expected to be used;
- global support if at all possible;
- transitions from the existing system to cause minimum operational disruption.

3.7.3.3 Issuer Identifier Numbers

Issuer identifier numbers are used for international telecommunication charge cards (ITU-T Recommendation E.118). An issuer identifier number identifies a national administration/ operator, allowing its customers to use their charge card in connection with various international services, at the appropriate charges for each transaction, and to have the charges billed to their account in the country that issued the charge card. These cards are issued in accordance with Recommendation E.118 and conform to the appropriate ISO standards jointly with the IEC, documented in ISO/IEC 7812 'Identification cards – Identification of Users', which are also used for credit cards.

3.7.4 BENEFITS AND DRAWBACKS OF A MOBILE TOP-LEVEL DOMAIN

With access to the Internet via mobile networks potentially being one of the main revenue drivers of future UMTS networks, the question of whether or not to use mobile-specific domain names has become an important and sometimes contradictory discussion point in the industry.

1. The UMTS Forum commissioned a study researching 'Benefits and drawbacks of a mobile top level domain (M-TLD)' (UMTS Forum, 2002c) with the support of Theron Business Consulting, Germany.
2. The UMTS Forum took the opportunity to start an initiative to pursue the application for new TLD names for future advantages.

There were two major sub-goals:

1. To describe strategic options around mobile-specific TLD names and assess the opportunities and risks with a focus on the marketing impact, while also looking at the technological implications, legal/regulatory requirements and internal organisation.
2. To develop reasonable alternative scenarios for introducing an M-TLD and to highlight the implications for these implementation scenarios.

This should help the UMTS Forum and its members to assess the efforts necessary and thus provide another criterion for deciding whether to go ahead or not with an application for a mobile-specific TLD. This sounds a little bit strange, but at the time of the research the industry was not completely convinced to continue with this idea. Eventually, there was a go-ahead much later, as will be pointed out at the end of this section.

3.7.4.1 Relevance of an M-TLD

The discussions with MNOs, ISPs, vendors and other market participants indicate that some key players generally have a positive attitude towards an M-TLD. However, within the group of MNOs, there is some reluctance to accept the benefits of M-TLD and a hesitation to accept the challenge a new TLD may offer. The following discussion will provide a better understanding of the topic and thus an answer to whether the challenge to introduce an M-TLD should be accepted.

There are two generic questions concerning the relevance of a TLD specifically for the mobile world (operators, services providers, vendors and users):

- For whom is it important?
- What are the drivers that make it important?

Mobile network operators are the focus of this study and will therefore be considered in the most detail. However, other market participants have also been examined with a view to understanding the implications they may face.

3.7.4.2 Potential Stakeholders

Mobile communications have been dominated by mobile network operators, who own the networks, provide the services and maintain the key customer relationships. In some countries, service providers play an important role in customer relationships. Recently, mobile virtual network operators (MVNO) entered the market and broadened this landscape. With the emergence of mobile access to the Internet, other players (e.g. content providers, portal owners) have also stepped in. With closer links developing between mobile networks and the Internet, it is expected that for the rollout of 3G all these forces will participate and help to develop the market and the revenue flows.

The M-TLD may be of benefit to all these market players, depending on the role each takes in the market and the tangible benefits offered by the TLD. Mobile markets are driven by three factors: technology, marketing and consumer behaviour. The study analysed these aspects and studied additionally the following scenarios:

- Scenario 1: Closed Shop. Mobile operators are exclusively entitled to use the M-TLD. Other market players are not qualified to use this M-TLD.
- Scenario 2: Open World. Everyone is entitled to use the M-TLD.
- Scenario 3: Code of Conduct. Users of an M-TLD have to accept a commonly agreed and mandatory code of conduct. Otherwise, the TLD is open to anyone.
- Scenario 4: Various M-TLDs. A number of TLDs for mobile access to the Internet and other services is available to all players and consumers.
- Scenario 5: Wait and see. The UMTS Forum does not take any action regarding an M-TLD. It will wait and see if other individuals or groups are pushing forward.

The essential findings are discussed below.

3.7.4.3 Findings

End-User Perspective

For the end-user added value is created with an M-TLD only if it is meaningful (discrete meaning), supported by many registrants (to establish enough awareness) and effectively content behind it is experienced by the customer according to the expectation. These three parameters are best met under the 'code of conduct' and 'closed world', scenarios assuming that closed world means not only MNOs but also other relevant market participants (MNO exclusive scenario not considered viable due to competition law issues).

Provider Perspective

For providers (all kind of providers, including vendors) basically the customer perspective must be taken to assess the different options. Additionally, competitive advantages have to be regarded. Under these two parameters, 'code of conduct' seems best. The alternative 'closed shop' in the sense of the .aero TLD (restricted to market players within air transportation) does not offer any additional benefits compared to 'code of conduct'. It is less attractive, due to the lack of control over content (type of content, quality of service etc.).

MNO Perspective

If MNOs could develop and maintain a major portion of the future mobile markets on their own (without any other players) the 'closed shop' approach would be ideal. However, first, it is commonly believed that all market participants will be required to develop the market and, second, competition law is a critical showstopper to this approach. Otherwise only the 'code of conduct' offers advantages of similar magnitude.

Overall

- The 'closed shop' would probably offer most advantages in terms of differentiation and value to be captured within the group of users. However, the contribution in developing the market will be quite limited. Competition law will probably block such an approach or demand other disliked obligations from MNOs.
- The 'code of conduct' seems much more appropriate for strategic (all players can participate, critical mass can be reached) and practical (no competition law issues, high likelihood of ICANN approval) reasons.
- No other scenario provides sufficient benefit to justify the efforts required to introduce an M-TLD.
- The key challenge for the 'code of conduct' scenario arises from its design – the rules and specification have to be mutually agreed. This will require a group of players (MNOs) to discuss their views and decide which product/service attributes and values should form the basis of a new M-TLD.

Given the expectation, that only 'code of conduct' is a viable concept to take forward, the implications related to an application and introduction of such a TLD have been investigated at very high level. No other scenarios have been considered in detail. The reason for this was the ambiguous position of the industry at that time, when one group was pushing forward and the other one was pulling back, which did not allow the Forum as a legal entity based on mutual consensus to pinpoint the received results in public to a more detailed degree.

To catch up on the final remarks above eventually the GSM Association took the results from the UMTS Forum's study and, in 2004, started, together with the industry, a request to ICANN to allocate a dedicated top-level domain for the mobile industry.

4

Technology

In this chapter a general view of some aspects of the technologies used in 3G is described. Other technologies such as speech coding, modulation schemes, protocols, etc. are described in detail in many other publications (Walke *et al.*, 2003) including standards, technical reports and guides of ITU-R, 3GPP and 3GPP2 . However, the reader will be given a short introduction to those technologies that are important in the context of this book. The given descriptions should be sufficient to understand its contents.

4.1 THE ITU IMT-2000 FAMILY CONCEPT

The original vision for 3G, as exemplified by the ITU's Future Public Land Mobile Telecommunications Systems (FPLMTS) concept (since renamed IMT-2000), envisaged the provision of tightly integrated voice and high data rate services to a handheld wireless terminal device. 3G was to be a global standard, facilitating international roaming. It was essentially a market-driven approach, conceived before the implementation of second-generation (2G) networks. Now that 2G networks have been deployed for more than a decade, the concept of 3G has had to be modified.

The success of 2G expanded the cellular market beyond all expectations. And the success of the Global System for Mobile communications (GSM) fundamentally altered the characteristics of that market (see Figure 4.1). GSM developed into a near-global system, providing widespread international roaming between all continents, and delivering both voice and data services to a handheld terminal. Within the large North American market, GSM was challenged by alternative 2G technologies in the form of time division multiple access (TDMA) (IS-136) and cdmaOne (IS-95). Japan introduced its own 2G standard, Pacific Digital Cellular (PDC), former Japanese Digital Cellular (JDC), as well as cdmaOne. Operators throughout the world invested heavily in 2G technologies to satisfy market demand.

Introducing 3G into this environment required a change of focus. The need to protect existing investments in different 2G technologies has shifted the drive toward a single global

The Mobile Multimedia Business: Requirements and Solutions   Bernd Eylert
© 2005 John Wiley & Sons, Ltd

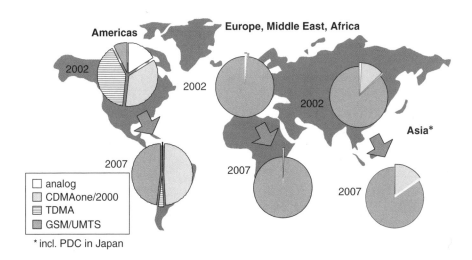

Figure 4.1 Subscriptions by regions and standards, 2002–2007 (estimated). *Source*: Data from Ovum and UMTS Forum (2002a).

standard. When you add the significant event of the emergence of the Internet, the additional capabilities of 3G became more focused on the provision of high data rates to deliver multimedia services. The emergence of the Internet as a mass-market content resource had justified the need for such high data rate capabilities and has since shifted the emphasis to packet-switched, Internet protocol (IP)-based core networks. There is general acceptance within the industry that 3G core networks will eventually be all-IP based.

The solution was the introduction of the IMT-2000 family of systems concept for 3G. One consequence of that solution is that a single global standard does not exist yet. However, the UMTS Forum and the Operators Harmonization Group (OHG) (a conglomerate of leading operators from Europe, the USA and Asia-Pacific, running GSM, TDMA-IS136 and cdmaOne-IS-95- systems) believe that progress of technology, operational deployments and market requirements will continue toward convergence. Another consequence – important when considering market perspectives – is that 3G now means different things in different parts of the world.

In Europe, 3G refers to the UMTS technology members of the IMT-2000 family, derived from GSM and deployed on new spectrum. There is a strong focus within the UMTS community on international roaming capabilities and the potential benefits of the economies of scale that result from a common standard deployed across many nations. The same UMTS technology standard will be used in South Korea, China, Japan and most of the Asian region. In Japan and South Korea, 3G means an opportunity to join the worldwide opportunity.

In the USA, 3G refers to derivatives of existing 2G technologies, deployed largely on occupied spectrum. 3G in the USA focuses more on high data rates; international roaming capabilities are not a significant concern. The USA has lagged behind other world regions in the deployment of 2G digital cellular (the majority of cellular subscribers in the USA are still on analogue systems). The USA will see on one hand an evolution of cdmaOne to cdma2000, and on the other hand a synchronised evolution from TDMA-IS136 and GSM throughout EDGE to UMTS/W-CDMA.

This situation reflects differences in culture and in the approach to standardisation. The large domestic market of the USA celebrates diversity. The US approach is to let the market determine which standards succeed. Much of the rest of the world takes the opposite approach, seeing standards as a way of creating markets. The success of GSM is a consequence of this approach. In this approach, the choice of the standard and underlying technology is itself the result of a competitive process.

The opposite approaches to standardisation will actually converge in the 3G environment following WRC-2000 in Istanbul, where spectrum was identified to enable 3G to deliver multimedia services. The results of WRC-2000 (see Section 5.3.6.) have opened up more potential markets for 3G. Where IMT-2000 is accepted, market forces will determine when the spectrum is released and what 3G technology will be deployed.

However, the market for 3G is radically different from that for earlier generations of cellular, which owe their success to the addition of mobility to voice communications. The market now takes mobility for granted. Simply adding mobility to data communications will no longer command a sufficient premium to justify the introduction of 3G.

3G will have to satisfy market expectations. Those market expectations are determined by developments in the fixed Internet world, which is now delivering high-speed access to content. Both the access and the content are perceived by consumers to be low, or even zero, cost. Market expectations for 3G today are for affordable multimedia communications to handheld and portable devices – delivering the same functionality as can be achieved in the fixed environment and at similar price levels. Providing such functionality at a price acceptable to the market can only be achieved through technologies with a sufficient global reach to achieve significant economies of scale.

The history of cellular illustrates that market success is determined more by the availability of terminals and transport at reasonable prices than by the provision of network resources. Terminal manufacturers face a multiplicity of choice in a multi-mode environment and naturally focus their resources on developing products for what they perceive to be the largest potential markets. Economies of scale rule supreme in cellular terminal manufacturing, which is rapidly becoming a consumer electronics environment.

Without massive economies of scale, technologies will not be able to survive in the 3G environment. Intermediate or regionally based systems will become, in the words of Herschel Shosteck Associates (1999), 'orphan technologies'. This applies not only to 3G technology candidates but also to some of the interim technologies being introduced on the road to 3G.

In 2G technologies, according to the annual database of UK analyst EMC, GSM has about two thirds of the world market. Japan has decided that its PDC 2G technology will not be evolved to 3G but will be replaced by the UMTS/IMT-2000 technologies. The TDMA and GSM communities are working on harmonisation procedures in the approach to 3G. The 15% of the world market currently using cdmaOne technology, mainly located in the USA and South Korea, have a transition path to the IMT-MC member of the IMT-2000 family, but are limited to existing spectrum.

Assuming that deployed 3G technologies will follow migration paths from 2G, it is clear that UMTS will enjoy significant economies of scale worldwide. In the USA, GSM is close to achieving a national footprint that should eventually ensure roaming capability with UMTS service providers worldwide using multi-band rather than multi-mode handsets – a much more attractive proposition for terminal manufacturers.[1]

---

[1] Multi-mode solutions such as GSM/DECT and intermediate technologies such as High Speed Circuit Switched Data (HSCSD) have been deployed by operators, but suitable terminals have failed to materialise in volume.

With a potential market size of less than 20%, cdma2000 has been forecast to become a 'semi-orphan technology, gradually falling behind IMT-2000/UMTS' (Herschel Shosteck Associates, 1999).

The same may not be true for the low chip rate TD-SCDMA (Chinese proposal for the TDD mode). The possibility of implementing IMT-TC rather than IMT-DS technology in the vast rural areas of China would immediately give this option sufficient economies of scale to be economically interesting. In addition, it will give Chinese industry some IPR and a better position in negotiations to accept other standards of the IMT-2000 family. Anyhow, by autumn 2004 the Chinese-developed TD-SCDMA 3G standard had been found to be lagging behind rival 3G wireless technologies, but eventually Chinese industry will become an important player in 3G technologies (FierceWireless, 2004b).

There is a belief in the industry that 3G will play a role in meeting the communication needs of developing countries, helping to narrow the digital divide surrounding access to the Internet that increasingly separates the developed from the developing world. In general, mobile communication is seen as very important for developing countries to build up economic activity. However, it is unlikely that 3G networks will have a significant deployment in most of the developing countries between now and 2010, because the 2G mass market in these countries is just beginning to start.

4.2 TECHNOLOGY COMPETITORS

There are a lot of other competing technologies for 3G/UMTS in the public and non-public environment (UMTS Forum, 2000b). There is a clear advantage for the end-user to be able to use the same terminal both indoors and outdoors, and it is expected that there will be applications that end-users can use both inside and outside buildings. The support of the same applications on the same (low-cost) terminal also ensures a consistent interface between applications and the end-user.

The following competing radio technologies should be noted (Huber and Huber, 2002):

- Infrared for very short range applications in one room (e.g. remote control, PC to mobile, one voice channel).
- Bluetooth for short range applications (e.g. wireless PC, optimal headset, communications between PC and electronic organiser, home automation).
- HomeRF/SWAP for short/middle range applications (voice and data application, WLAN).
- 802.11x family of IEEE wireless standards for short/middle range applications (WLAN, cellular corporate networks on campus for voice and data services up to 54 Mbps for SOHOs and SMEs).
- HIPERLAN/2 for short/middle range applications (multimedia with bandwidth requirements up to 25 Mbps and requirements for QoS support (high end applications), WLAN).
- DECT for short/middle range application (voice and data application, VoIP, WLAN, cellular W-PABX/corporate network on campus for voice and data services up to 2 Mbps for SOHOs, SMEs).

Table 4.1 compares the advantages and limitations of these technologies.

Table 4.1 Comparison of different wireless technologies

Technology	Cellular	ISM band	Effective cell radius (m)	Level of penetration		Advantages	Limitations (today)
				Residential	Corporate		
Infrared	No	No		High	Low	Price	One room, line of sight
Bluetooth	No	Yes	10	Low	High	Nearly all applications in a sphere of 10 m, ISM band	<1 Mbps, 8 active devices
HomeRF/ SWAP	Yes	Yes	<50	High	Low	Combines benefits of 802.11 and DECT	Smaller cells due to electronic smog
802.11x	Yes	Yes	<50	Low	High	Best wireless data for SME; high rate – up to 54 Mbps in ISM band	Electronic smog/lack of interoperability
HIPERLAN/2	Yes	No	<50	Low	High	Max. 25 Mbps in ISM band; QoS support	Price range (high)
DECT	Yes	No	20–300[a]	High	High	Proven technique; IMT-2000 standard	Protected band only in EU
Licence exempt UMTS (3G)	Yes	No	20–300[a]	High	High	One technology at any place; protected IMT-2000 band	Not allocated worldwide

[a] Not indoors.
© UMTS Forum.

The weakest points of all technologies operating in the Industrial, Scientific and Medical (ISM) band are security and the electronic 'pollution' from the many other uncontrolled sources of RF energy already existing in this band and likely to be added to it, causing a high level of interference. This can dramatically reduce the effective transmitted bit-rate and the cell radius. Some theoretical advantages of these different competing technologies as shown in the table are therefore significantly reduced. Today only the low-end 2.45 GHz band seems to approaching congestion; in the 5 GHz band used by HIPERLAN/2 (Europe), MMAC (Japan) and IEEE 802.11a and g (USA) such problems are not expected in the medium term due to an initially higher price level of terminal devices. For more information, see Section 4.7.

4.3 IMPACT ON STANDARDISATION

Wireless access has already introduced a new set of standards and protocols that add a layer of complication to application solutions which are not necessarily compatible with the Internet world (UMTS Forum, 2000b). This experience was shown with the introduction of voice/video over IP, WAP, I-mode and similar IP services to mobile networks.

Figures 4.2 and 4.3 show the main difference in the standardisation of IMT-2000 in comparison to UMTS in the focus of an extended vision. Figure 4.2 shows the involved standards in the IMT-2000 scenario linked with the ITU-based fixed network environment. A series of network services are specified entirely according to the ITU framework, including service interworking, addressing, signalling and QoS standards.

For UMTS these standards have to be supplemented for those services which will also be offered also on the wireline network, e.g. bitrate harmonisation, circuit switched service interworking etc. As an example, UMTS provides 384/128 kbps transparent connections, whereas the fixed ISDN network only offers 64 kbps and 128 kbps in a limited way. Also the harmonisation of the terminal interworking characteristics between wireless and wireline terminals may be a standardisation issue, e.g. for video telephony, VoIP, fax and other telematic services. QoS, security and billing issues could also have an impact on standardisation.

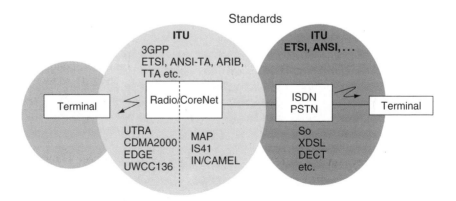

Figure 4.2 UMTS standardisation scenario in the framework of IMT-2000. © UMTS Forum (2000c).

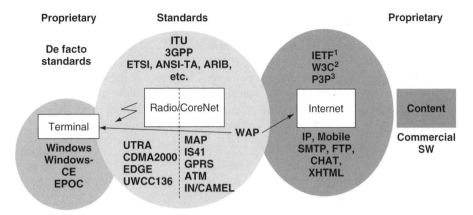

[1]IETF = Internet Engineering Task Force   [2]W3C = World Wide Web Consortium   [3]P3P = Platform of Privacy Preferences
[4]XHTML = an XML version of HTML (new mark-up language recommended by W3C in early 2000)

Figure 4.3   Mobile multimedia portal platform. © UMTS Forum (2000c).

Figure 4.3 shows the scenario for multimedia services: IMT-2000/UMTS is linked with the Internet, and through portals to the provision of content, and is therefore faced with a number of new interworking issues. They include quality, security, mobility management, billing etc. A main issue is the interworking on the protocol layers, where out-of-band and in-band control functions have to be aligned. The standardisation also has to specify impacts regarding addressing (ITU, IP), which is quite different in IPv4 and IPv6. The transition phase from IPv4 to IPv6 is not only impacting UMTS itself, but also the worldwide Internet and intranets. Thus, this topic will play a significant role in the future standardisation within 3GPP as well as IETF.

Harmonisation of the UMTS standards in the IMT-2000 framework with the standards on the Internet side is necessary to make the mobile multimedia portal platform a workable solution in an international networking environment, especially for the roaming user. Especially for this case, email standards have to be specified based upon the existing IETF SMTP protocol standard. Regarding the portals, HTTP standard issues need to be discussed in the relevant bodies to guarantee for the end-user XHTML/XML transparency and WAP compatibility.

The result is a widening of the scope in the standardisation activities. Also important is the backward compatibility with 2G systems, especially with GSM and the interworking with WAP. These subjects will need to be addressed by the 3GPP, IETF, IPv6 Forum etc.

4.3.1   MOBILE MULTIMEDIA PORTAL PLATFORM

The mobile multimedia portal (MMP) platform is located at the transition between the access and transport network and the content provision. The MMP deals with the mobile user on the applications level, on top of the relevant protocols like HTTP etc. Thus, the question arises in what way the MMP needs to be standardised and to what extent, e.g. in the areas of applications-dependent QoS, security and billing. Then there is the choice of languages such as HTML, XML and WML. There must also be a common set of rules on portal functionality, especially important for the roaming user.

In the following sections the two main topics of *mobility* and *security*, which will impact standardization, are discussed. In 3GPP, UMTS Release 99 inter alia deals with two separate UMTS core network domains, namely a circuit switched domain based on GSM Mobile Switching Centers (MSCs), and a packet domain built upon GPRS Support Nodes (GSNs). UMTS Release 5 defines features for further evolution following GSM/Internet convergence.

4.3.2 MOBILITY

Mobility, in all its many forms, is becoming the watchword of our society. Everything moves faster and faster. IP seems to be the end-to-end protocol for the future delivery of most services since it will exist in the wireline and wireless worlds, in office extension environments and in home networks.

The increasing interest in mobile IP as a potential mobility solution for cellular networks leads to new solutions and extensions to the existing protocol world in telecommunication networks. There is also a need to put together the demands on mobile IP, from a cellular perspective, in order to harmonize the evolution of mobile IP and the existing mobility solutions in cellular networks. As a network-layer protocol, mobile IP is completely independent of the media over which it runs, i.e. independent of technology. This is in line with the Internet protocol philosophy itself, which was designed to be independent of the underlying characteristics of the links over which it runs.

Mobile IPv6 as standardised by IETF benefits today by being an integral part of the ongoing development of the Internet and cellular standards. It allows the user to keep his/her home address while roaming and remain in a state of 'always on'.

IPv6 mobility determined by and optimised for mobile terminals will be one of the major features in cellular networks. IPv6 provides maximal and transparent interoperability with core networks. It is simple and scalable and can support billions of users' new devices. Its key benefits are mobility, security, QoS, scalability and auto-configuration. But it also offers 'built-in' IP Security (IPSec, a protocol task group of IETF), which provides security management for applications. It supports cellular and non-cellullar access. The flexibility allows sharing of resources using a diversity of technologies, both wireline and wireless. It has no impact on location registers since the information required to route packets is managed independently of the information used to locate and authenticate a UMTS user.

The IETF is currently working with 3GPP to incorporate the UMTS requirements on terminal mobility into mobile IPv6 to provide seamless wireless and wireline mobility management. This will offer a major advantage in comparison to second-generation cellular network roaming concepts (optimized routing).

Each mobile node is always identified by its home address, regardless of its current point of attachment to the Internet. While situated away from its home, a mobile node is also associated with a care-of address, which provides information about the mobile node's current location. IPv6 packets addressed to a mobile node's home address are transparently routed to its care-of address. The protocol enables IPv6 nodes to cache the binding of a mobile node's home address with its care-of address, and then to send any packets destined for the mobile node directly to it at this care-of address. The way this is done is as follows:

- advertisement from local router;
- seamless roaming: mobile node keeps home address;
- address auto-configuration for care-of address;
- binding updates sent to correspondent nodes;
- mobile node 'always on' by way of home agent.

Figure 4.4 illustrates one possibility for the start-up phase of UMTS to solve the provision of mobility control functionality across heterogeneous networks. Figure 4.4 presents a first option in providing mobility functionality over heterogeneous networks; that is, an interworking solution: within the UMTS system, classical MAP-based mobility functions are exploited, while on the IP side, mobile IP can be used. In this option a complete interworking function has to be developed in order to allow the user overall mobility across both the mobile and the IP environment.

Figure 4.5 illustrates a transparent IP mobility solution: both within the UMTS system and on the IP environment, and will be implemented after the IETF/3GPP solution is available as unique mobility support functionality. In this case a transparent end-to-end mobility protocol

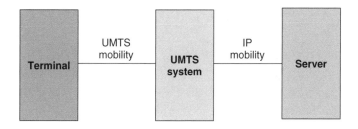

Figure 4.4 Mobility interworking. © UMTS Forum (2000c).

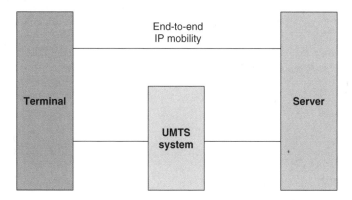

Figure 4.5 Transparent IP mobility. © UMTS Forum (2000c).

is provided through the exploitation of IP integrated functions, i.e. mobile IP for the discrete macro-mobility, within and between core networks, and cellular IP for the continuous micro-mobility, within the access network.

4.3.3 SECURITY

Many of the existing infrastructure elements in mobile networks, such as switches, home location registers and authentication centers, will be duplicated in the IMT-2000/UMTS networks. These will be connected to the Internet, which generally has network redundancy rather than duplicated network elements. This has to be taken into consideration in cases where UMTS operators build their own IP infrastructure. Additionally, external interfaces to the public Internet will have to be protected by methods such as firewalls or blocking communication coming from unknown destinations. Attacks on networks can be expected to increase as networks are opened up to Internet hacker communities, because the hardware used will be based on well-known Internet technology, rather than telecommunications networks and switches. Furthermore, sending control signalling over the Internet is more than likely to introduce further security threats.

Conventional fraud detection systems, which allow fraud to be identified from switch records (which show information such as call date, time, destination, type) will no longer work. Such information, as well as the customer communication, may well be hidden from the operator, using techniques such as VPN or IP tunnelling (see Section 4.5.1.2).

4.3.3.1 Billing and Security Infrastructure

Billing and the ability to authenticate and authorise users to access the public IP-based mobile network are crucial features that must be available as part of an all-IP solution for a public 3G system. Mobile IP either does not provide this functionality (billing, authorisation), or does not provide it adequately (security).

The Internet Engineering Task Force (IETF) is currently working on the specification of a generic Authentication, Authorisation and Accounting (AAA) protocol that would provide this type of functionality.

In order to meet the security requirements of the 3GPP/3GPP2 networks there is a require-ment to resolve issues in two primary areas, AAA services and mobile IP. AAA is required for authentication, authorisation and billing. The remaining issues centre around cross-domain AAA, authentication using Public Key Infrastructure (PKI). At the moment there is considerable aversion to use of IP Security and Internet Key Exchange (IKE) protocols due to perceived overhead and delay.

4.3.3.2 Encryption

3GPP decided to use for UMTS an algorithm for confidentiality and integrity protection of commands, called KASUMI, based on the Japanese algorithm MISTY1. Work was inter alia funded by the European Union's research project ACTS as 'UMTS Security Architecture' (USECA). Encryption standards and security architectures (TS 33 series and TS 35 series) are published on its Internet site for public scrutiny. Authentication is similar to GSM, allowing choice of algorithms. The network is authenticated to the mobile to minimise the false base station attack, possible in other technologies, by using the Message Authentication Code (MAC).

Encryption standards and security architecture have been adopted by the American National Standardization Institute, Technical Regulation Group 45 (ANSI-TR45) (as a member of 3GPP) for North America, which will allow roaming from the security architecture point of view between the different IMT-2000 technologies, as well as backward compatibility with GSM, ANSI-41 and ANSI-136. The reader can find an extensible description of mobile security on 3GPP's website (www.3gpp.org) and in Hillebrand (2002).

4.3.3.3 Technical Fraud

Both the terminals and the network itself will be connected to the Internet. This means that they must be protected by firewalls, otherwise they will be accessible by others connected to the Internet.

Terminals will be capable of running Internet browsers to access information, as well as making voice calls or recording pictures. This implies that the terminals will be capable of executing programs, in the same way as Web browsers such as Internet Explorer or Netscape do at the moment. So, there is a possibility of viruses or Trojan horses (rogue code) in the mobiles. These could, for example, make calls without customers being aware, seize telephone directories, find the location of the mobiles or even modify financial transfers in progress. Although security and protection exist within the standards, you have seen what happens with security on the Internet with complex systems.

As well as the normal identities, such as a telephone number, the terminals will also be known by Internet addresses, which may be fixed or changed for every call. Unfortunately, although the network will know the connection between these, this information will probably not be known to other users on the Internet. This will mean that there is likely to be more protection of the customer and anonymity, but it will be harder to identify people and they will have multiple identities.

4.3.3.4 IP Security

Rapid advances in communication technology have accentuated the need for security on the Internet. IPSec has developed mechanisms to protect client protocols of IP. A security protocol in the network layer is developed to provide cryptographic security services that will flexibly support combinations of authentication, integrity, access control and confidentiality. The protocol consist of three core components:

- The IP Authentication Header (AH), which verifies the identity of a packet's sender and the authenticity of the packet's contents.
- The IP Encapsulating Security Payload (ESP), which encrypts a packet before transmitting it and may also encapsulate the original IP packet. It is independent of the cryptographic algorithm.
- The Internet key exchange (IKE), which governs the transfer of security keys between senders and receivers.

The preliminary goals will specifically pursue host-to-host security followed by subnet-to-subnet and host-to-subnet topologies. AH and ESP can be used with various authentication and encryption schemes, some of which are mandatory.

Protocol and cryptographic techniques have also been developed to support the key management requirements of the network layer security. The Internet Key Management Protocol (IKMP) will be specified as an application layer protocol that is independent of the lower layer security protocol.

IPSec gateways work only in tunnel mode, which means no part of the original packet is vulnerable to interception. Some of the key IPSec features are:

- Signalling: integrity, authentication, anti-replay protection
- User traffic: integrity, authentication, confidentiality
- Visited network resources and traffic: access control, confidentiality
- No foreign agents
- Integration: IPSec protocols into IPv6 devices as standard
- Elimination: home address option from network-ingress filter problems.

Route-optimisation functionality is integral. Similarly to the mobility problem of providing complete and reliable end-to-end functionality across heterogeneous networks, security will follow the same approach.

3GPP is performing a sanity check of IPv6 and the different add-on components, i.e. MIP, IPSec etc. It is expected that the necessary amendments meeting the requirements of a huge cellular market will be made where appropriate.

Security relationships (Figure 4.6) in a heterogeneous network environment can be very complicated, actually requiring both an interworking and a transparent end-to-end solution. In fact, dedicated mutual authentication and encryption facilities are used to check and secure the transmission segments in each specific environment, i.e. both the UMTS system and the IP segment. While end-to-end transparent solutions are needed, the application layer will be secured via e.g. digital signatures plus non-repudiation techniques. Non-repudiation can be achieved by generating a signature and combining it with some form of user authentication data (such as a PIN). When the user commits to transaction, the signature will ensure that he/she cannot deny later that it took place. Of course, this kind of functionality is fundamental in any kind of mobile commerce transaction.

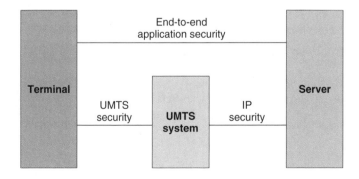

Figure 4.6   Security relationships. © UMTS Forum (2000c).

In the security domain global coordination is necessary between:

1. the two different solutions adopted at the transport layer, which must interwork in an efficient way;
2. the solutions adopted at the different layers, i.e. transport and application, which together must guarantee the required level of security.

4.3.4 FURTHER ASPECTS

While the fundamental issues are being actively addressed, some key areas are now being tackled. The major issues are listed below. The IDPRs from the transport and content layers must be linked in order to ensure the correct accounting can take place. Pre-paid will demand real-time metering, or real-time mediation methodologies, and is complicated by the uncertainty of estimating how long a session will last. It will not be acceptable to cut off a session during a piece of music or video.

Roaming will require both the delivery of personalised services while on a visited network, but also localised services, such as directory type services. Pre-paid while roaming will need to be addressed. The GSM standards of TAP3 and CAMEL provide the accepted building blocks to address this issue.

Quality of service issues are being addressed, and a dialogue between billing organisation, mediation companies and network equipment manufacturers is being formalised. Interconnect agreements will become more complex when content delivery and accounting are considered. Access to and usage of customer data needs careful attention, especially in light of national regulation.

The impact of IP addresses is another issue to be observed. The current version of the Internet protocol is IP version 4 (IPv4). In this protocol, the IP address of the mobile machine does not change when it moves from a home network to a foreign network. In order to maintain connections between the mobile node and the rest of the network, a forwarding routine is implemented. When a person in the physical world moves, they let their home post office know to which remote post office their mail should be forwarded. When the person arrives at their new residence, they register with the new post office. This same operation happens in mobile IP. When the mobile agent moves from its home network to a foreign (visited) network, the mobile agent tells a home agent on the home network to which foreign agent their packets should be forwarded. In addition, the mobile agent registers itself with that foreign agent on the foreign network. Thus, all packets intended for the mobile agent are forwarded by the home agent to the foreign agent, which sends them to the mobile agent on the foreign network. When the mobile agent returns to its original network, it informs both agents (home and foreign) that the original configuration has been restored. No one outside the networks needs to know that the mobile agent moved. This configuration has been widely implemented and works in general quite well, but it has some drawbacks. Depending on how far the mobile agent moves, there may need to be some store and forwarding of packets while the mobile agent is on neither the home nor the foreign network. In addition, mobile IP works only for IPv4 and does not take advantage of the features of the newer IPv6. However, industry and the IPv6 Forum are working together to make this protocol usable not only in the wired world but also in the mobile world. This is very important because with the installation of GPRS and UMTS in the networks a shift from Ipv4 to Ipv6 could be done more easily now than later. A full explanation of the IP addressing scheme is given in Section 4.5.4.

4.4 TERMINALS AND DEVICES

The merging of the mobile phone and the personal organiser has been promoted as the obvious executive toy (UMTS Forum, 2000b). However, no single type of device will win out in the collision of technologies. There will be a divergence in philosophies of mobile data, and a wide range of devices will compete with one another, and may also complement one another, e.g. mobile phones incorporating MP3 players for downloading music from the Internet will appeal to the young, while many hard-pressed executives will still want to see a screen full of data on a notebook. Technologies such as Bluetooth that enable wireless communication between portable devices mean that consumers will be able to carry around phones and other devices such as palmtops or organisers that communicate with one another rather than one do-it-all handheld.

So mobile terminals will allow people to surf the Internet, rapidly download emails, music and high quality pictures, and even hold videoconferences on the move. Already today the latest generation of Web-enabled cell phones seems to have all the functions business people need, e.g. alarms, calculators, calendars, email, e-wallet games and phone directories.

The multimedia drive is also likely to intensify once broadband Internet access arrives. Terminals will need increasing amounts of computer power so they can handle full-screen, high-quality video content. And when it comes to Web sites, technologies which insert video clips into Web pages are just the beginning and icons on the display will be pictures/graphics.

Delivery of multimedia data depends on standards defining format (and compression) and how transmission should be enabled. For pre-generated images, JPEG (Joint Photographic Experts Group), GIF (graphics interchange format) and DIB (device-independent bitmap) are among the best known. For pre-generated audio, WAV and MIDI (music instrument digital interface) are widely used. Then there is MPEG-1 (Motion Picture Experts Group) Layer 3, popularly known as MP3, which stores almost CD quality sound at compression ratios of up to 10:1. And a popular format for pre-generated video is Video for Windows (kept in files with AVI, or Audio Video Interleave, extension) from Microsoft Corp. For very high quality video, whether it is pre-generated or streaming, MPEG-2 and MPEG-4[2] can be used.

A decade from now, the Internet will look and behave completely differently. Accordingly, the content availability and Internet will have a strong influence on the devices as well.

The semiconductor industry is already building whole systems on a chip. These multimedia-enabled systems will be made small enough to be used in Web pads, mobile phones, thin clients, set-top boxes, and a host of other, yet-to-be-conceived devices. They will offer the flexibility to be used in any way required. It is most likely that the multi-mode, multi-band and multi-standard issues will be solved by soft-defined radio concepts in the future.

Figure 4.7 illustrates the convergence of different types of well-known equipment functionality towards the fully integrated multimedia terminals of the future.

---

[2] MPEG is ISO/IEC JTC1 SC29 WG11's Motion Pictures Expert Group; the standard MPEG-4 can be used from 15 kbps up to around 50 Mbps depending on application.

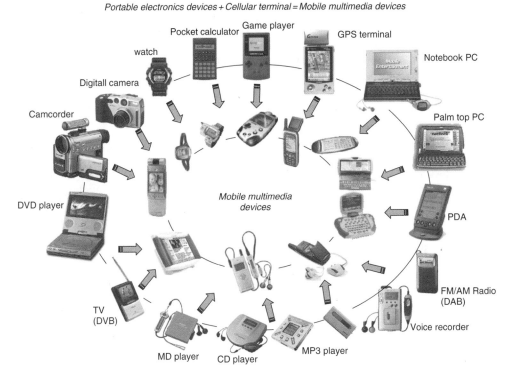

Figure 4.7 Multimedia devices. *Source*: Alps Electric, March 2000. © UMTS Forum (2000b).

4.5 NAMES AND ADDRESSES USED IN GPRS

4.5.1 ACCESS

4.5.1.1 Addressing and Routing Currently Used with GPRS

The UMTS architecture will be based on that developed and currently being brought into service for GPRS. This architecture introduces several concepts that need to be distinguished carefully to understand the naming and addressing requirements.

The GPRS networks are designed to provide access links between mobile terminals and the terminal's home IP network. The home IP network may be:

- an ISP
- a corporate IP network, e.g. an Intranet.

In practice, a GPRS operator may also own and run an ISP that is connected to the GPRS network, but the following explanation treats these two functions separately.

4.5.1.2 The Access Tunnel across GPRS

When GPRS networks are interconnected and provide roaming, this will mean that a terminal has worldwide access to its ISP. The access provided by GPRS is analogous to a dial-up connection over the PSTN to an ISP or a leased line connection to an ISP. The access link has the form of a tunnel through the GPRS networks. The arrangement is shown in Figure 4.8.

The mobile terminal will be assigned an IP address by the ISP (not by the home GPRS network). This address may be public or private and may be assigned permanently or temporarily. In Figure 4.8 the terminal has the address jkl.mno.pqr.stu from the range jkl.mno.pqr allocated to the ISP. This IP address is not seen by the GPRS networks as it passes unexamined through the tunnel.

The names used by the mobile terminal (or its user) will be specific to the service or services supported. These names will be related to the ISP or Intranet and may also not be known to the GPRS network. For many services a name of the form user@domain will be used. The 'domain' may be the identity of the ISP (e.g. for an individual subscriber) or a domain name that is served by the ISP but is in principle portable to another ISP. In Figure 4.8 the user has the name abc@xyz.com from the ISP or corporate network xyz.com.

Thus the mobile terminal always appears to the outside world, e.g. the Internet, as part of the ISP or intranet, even though it may be on any GPRS network, either its home network or roaming.

The mobile terminal is able to select the ISP or intranet that it accesses and may have different names on different ISPs (similar to a user logging into different email accounts).

4.5.1.3 The Structure within GPRS

GPRS networks have a structure similar to GSM networks:

- The Serving GPRS Support Node (SGSN) is equivalent to the MSC and connects to the radio base station controller and then to radio base stations in an area.
- The Gateway GPRS Support Node (GGSN) is equivalent to the GMSC and provides the connection point to other networks.

The GPRS nodes are interconnected on an IP network and the IP networks of each GPRS network are interconnected on an inter-PLMN IP backbone. The Inter-PLMN backbone is at present an isolated IP network and not connected to the public Internet. However, the GSM operators

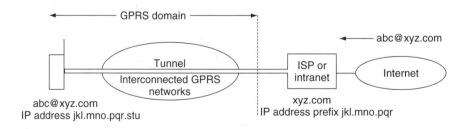

Figure 4.8 The GPRS tunnel. © UMTS Forum (2001c).

have arranged to use public IP addresses for interfaces on this network to ensure uniqueness and forward compatibility (if ever this network is connected to the public Internet). These IP addresses are used to identify interfaces on the SGSNs and GGSNs.

The tunnel through the inter-PLMN backbone between the SGSN serving the mobile and the GGSN is established by the GPRS Tunnelling Protocol (GTP).

When a GPRS terminal logs on to its home or a roaming network, it selects which ISP/ intranet it wishes to access. This is done by specifying the Access Point Name (APN) of the ISP/intranet. Access may go through the visited GGSN or the home GGSN depending on the connections available; so, for example, a roaming terminal can access a local node of its ISP via the visited GGSN. The following example is for a roaming terminal accessing its ISP via the home GGSN:

1. The user identifies the requested ISP as xyz.com and the terminal sends this information to the SGSN. The network ID part of the APN (xyz.com) is a private GPRS equivalent. However, it has been agreed that operators should only allow public registered domain names to be registered. For connections without public registered domain names the operator could use its own domain name, e.g. smallcompany.t-mobile.de, anothercompany.t-mobile.de etc.
2. The SGSN adds the operator ID of the visited network to give xyz.com.mnc01.mcc007.gprs, where the visited operator ID is mnc01.mcc007, (mnc=mobile network code; mcc=mobile country code). This string is the whole APN.
3. The visited network interrogates its own DNS (see Section 4.5.2) but the interrogation fails as xyz.com is not known in the roaming network.
4. The SGSN changes the operator ID to the home operator and sends the request to the root DNS (not one of the 13 public root DNS servers but a private server exclusive to the mobile networks). The root server forwards the request to the home network DNS.
5. The home DNS identifies the IP address of the GGSN that serves the requested APN.
6. The SGSN establishes a tunnel to the GGSN using the GTP and the GGSN establishes the link to the ISP.

The sequence of interrogations is shown in Figure 4.9. The tunnel is connection oriented and is held as long as the terminal remains logged on. It has a tunnel identifier (TID). It extends from the mobile to the ISP.

For the section of the tunnel between the SGSN and the GGSN, i.e. the section across the inter-PLMN backbone, there is a tunnel identity for each user to identify that user's tunnel

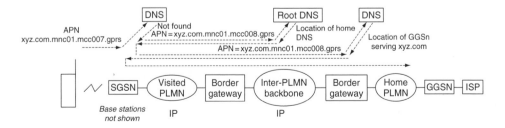

Figure 4.9   Sequence of interrogations for an outgoing call. © UMTS Forum (2001c).

ID, which relates to the GTP protocol running between the SGSN and the GGSN, and there are IP addresses for the source and destination SGSN/GGSN interfaces.

Figure 4.10 shows communications in progress between a mobile and an entity in the public Internet. Packets pass transparently through the tunnel between A and B with the public IP addresses of A and B. The address of A was allocated by the ISP.

In the section between the visited SGSN and the home GGSN, these packets are encapsulated (wrapped up) in the following additional headers and footers in the following order, working outwards:

- Tunnel ID from the GTP;
- UDP port to identify that the GTP is used from the UDP protocol that supports GTP;
- IP addresses for the visited SGSN and home GGSN.

This is shown in Figure 4.11.

4.5.2 DOMAIN NAMES

Top level domains were introduced briefly in Section 3.7.1.2 in the context of 'Naming'. To give the reader a quick reminder, TLDs may be either open to any use or limited to a specific use

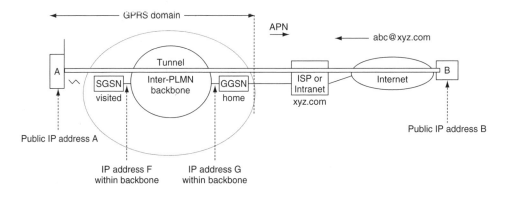

Figure 4.10   Naming and addressing for the tunnel. © UMTS Forum (2001c).

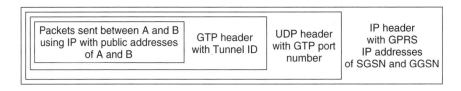

Figure 4.11   Encapsulation of headers within a tunnel. © UMTS Forum (2001c).

or charter. In practice, although .com, .org and .net were intended for specific applications (chartered) they have become open to any use. The same is true for some country codes, which may be used by organisations outside the boundaries of the country concerned.

ICANN decides whether, how and when to add new gTLDs to the domain name system. A number of plans have been proposed to create new gTLDs, such as .firm, .store, .law and .arts, and some companies have even taken orders for them. ICANN commenced a formal process of inviting proposals for new TLDs in August 2000 and decisions on the allocation of new TLDs have been made. Examples are .aero, .bank and .info.

According to ICANN, there are many arguments both for and against new gTLDs: for example, those in favour argue that new gTLDs are technically easy to create, will help relieve perceived scarcities in existing name spaces, and are consistent with a general push towards consumer choice and diversity of options; those opposed point to greater possibilities for consumer confusion, the risk of increased trademark infringement, cybersquatting (use of a name without payment) and cyberpiracy (taking someone else's traffic). In practice, many large companies with valuable names will register their names under all available TLDs in order to protect their identity.

Unlike E.164, the domain name system is supported by a global domain name server system (DNS) that resolves domain names into IP addresses, which are then used for routing Internet traffic. DNS is a structured system of servers, at the top of which are the root servers.

ICANN handles the management of the root server system, which consists of a set of 13 file servers, which together contain authoritative databases listing all TLDs. Currently, the US Department of Commerce and Network Solutions Inc. (NSI), under an agreement with ICANN and the US Department of Commerce, operates the 'A' root server, which maintains the authoritative root database and replicates changes to the other root servers on a daily basis. Different organisations, including NSI, operate the other 12 root servers (including Lynx in the UK, which runs a server for RIPE, and a server in Stockholm). The US Government plays a role in the operation of about half of the Internet's root servers.

The function of the root servers is to resolve a solution from the TLD name to an IP address by which a TLD server can be contacted. The TLD server then resolves the second level domain (SLD) name into an IP address by which an SLD server can be contacted. Figure 4.12 shows the sequence for resolving the name xyz.co.uk into an IP address.

A master registry is maintained for each TLD name. Allocations of SLD names are made by registrars who update the registry.

For .gov, .edu, .mil and .int there are only single registrars, but for .com, .net and .org there are now competing registrars who issue SLD names. The system was changed in 1999 from single (monopoly) registrars to the new shared registry system in response to proposals from the US Government in its Green and White papers.

The registry for .com, .net and .org was run by InterNIC®, which was replaced by ICANN in 1999. Network Solutions is the largest registrar and was formerly the only registrar. There is now a rapidly increasing number of accredited and operational registrars. The process of accreditation is handled by ICANN and takes up to 30 days.

ICANN has published a proforma agreement that has to be entered into by registrars. Figure 4.13 shows the flow of the process of obtaining a domain name either via an independent registrar or an ISP and having the domain name–IP address pair registered in the domain name server run by the registry.

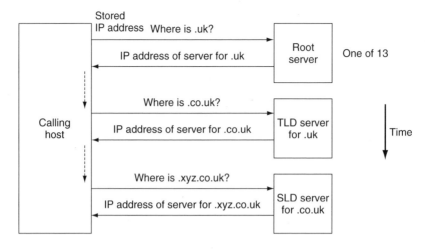

Figure 4.12   Example of domain name resolution. © UMTS Forum (2001c).

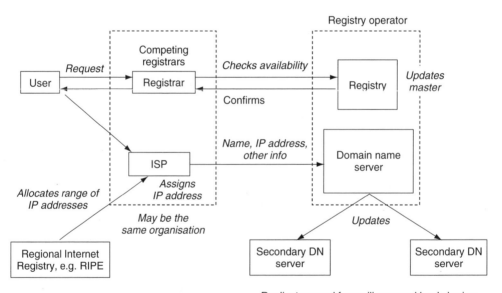

Figure 4.13   Registrar, registry and DNS functions. © UMTS Forum (2001c).

The DNS resolves domain names to IP addresses. There is a one-to-one relationship between domain names and IP addresses and so provision has been made for the reverse process, from IP address to domain name. However, because the DNS servers are structured by domain name, it is not possible to know which server holds the record with the IP address used for a reverse query. Therefore a special pseudo-domain has been created called in-addr.arpa. An IP address is stored in reverse form as this type of domain name. For example:

193.3.20.100 is stored as the domain name 100.20.3.193.in-addr.arpa

To enable reverse mapping, the assigned IP address has to be registered under the in-addr. arpa domain. The in-addr.arpa name space is divided according to the reverse of the IP addresses. Therefore 193.in-addr.arpa is delegated to the owner of the addresses starting with 193 and this organisation is responsible for giving the real domain name that corresponds to the address. Thus the name space under in-addr.arpa is structured according to addresses rather than the real domain names.

Reverse mapping is likely to be important for lawful interception.

4.5.3 ACCESS POINT NAME

In the GPRS backbone, an APN is a reference to a GGSN.[3] To support inter-PLMN roaming, the internal GPRS DNS functionality is used to translate the APN into the IP address of the GGSN.

4.5.3.1 Structure of APN

The APN is composed of two parts:

- The APN Network Identifier, which defines to which external network the GGSN is connected and optionally a requested service by the MS. This part of the APN is mandatory.
- The APN Operator Identifier, which defines in which PLMN GPRS backbone the GGSN is located. This part of the APN is optional.

The APN Operator Identifier is placed after the APN Network Identifier. An APN consisting of both the Network Identifier and Operator Identifier corresponds to a DNS name of a GGSN and has a maximum length of 100 octets.

The syntax of the APN must follow the Name Syntax defined in RFC 2181 (Clarifications to the DNS specification) and RFC 1035 (Domain names – implementation and specification). The APN consists of one or more labels. Each label is coded as one octet length field followed by that number of octets coded as 8-bit ASCII characters. Following RFC 1035 the labels should consist only of the alphabetic characters (A–Z and a–z), digits (0–9) and the dash (–). The case of alphabetic characters is not significant. The APN is not terminated by a length byte of zero.[4]

For the purpose of presentation, an APN is usually displayed as a string in which the labels are separated by dots (e.g. 'Label1.Label2.Label3').

---

[3] This description is based on 3GPP TS 23.003 V3.5.0 (2000–06) Release 99.

[4] A length byte of zero is added by the SGSN at the end of the APN before interrogating a DNS server.

4.5.3.2 Format of APN Network Identifier

The APN Network Identifier must contain at least one label and has a maximum length of 63 octets. An APN Network Identifier must not start with the strings rac, lac, sgsn or mc and it must not end in .gprs. It must not take the value '*'.

In order to guarantee uniqueness of APN Network Identifier within the GPRS PLMN(s), an APN Network Identifier containing more than one label corresponds to an Internet domain name. This name should only be allocated by the PLMN to an organisation that has officially reserved this name in the Internet domain. Other types of APN Network Identifiers are not guaranteed to be unique within the GPRS PLMN(s).

An APN Network Identifier may be used to access a service associated with a GGSN. This may be achieved by defining:

- an APN that corresponds to a DNS name of a GGSN and is locally interpreted by the GGSN as a request for a specific service, or
- an APN Network Identifier consisting of three or more labels and starting with a Reserved Service Label, or an APN Network Identifier consisting of a Reserved Service Label alone, which indicates a GGSN by the nature of the requested service. Reserved Service Labels and the corresponding services have to be agreed among operators.

4.5.3.3 Format of APN Operator Identifier

The APN Operator Identifier is composed of three labels. The last label must be 'gprs'. The first and second labels together uniquely identify the GPRS PLMN (e.g. <operator-name>. <operator-group>.gprs).

For each operator, there is a default APN Operator Identifier (OI) (i.e. domain name). This default APN OI is derived from the IMSI as follows:

mnc<MNC>.mcc<MCC>.gprs

where 'mnc' and 'mcc' serve as invariable identifiers for the following digits, and <MNC> and <MCC> are derived from the components of the IMSI defined in subclause 2.2.

This default APN Operator Identifier is used in inter-PLMN roaming situations when attempting to translate an APN consisting of Network Identifier only into the IP address of the GGSN residing in the HPLMN. The PLMN may provide DNS translations for other, more human-readable, APN Operator Identifiers in addition to the default Operator Identifier described above.

In order to guarantee inter-PLMN DNS translation, the <MNC> and <MCC> coding to be used in the mnc<MNC>.mcc<MCC>.gprs format of the APN OI shall be:

- <MNC> = 3 digits
- <MCC> = 3 digits
- If there are fewer than three significant digits in MNC, one or more '0' digit(s) are inserted at the left-hand side to fill the three-digit coding of MNC in the APN OI.

For example, the APN OI for MCC 345 and MNC 12 would be coded in the DNS as mnc012. mcc345.gprs.

4.5.4 IP ADDRESSES AND THEIR ALLOCATION

4.5.4.1 IPv4 Addresses

IPv4 uses a 32-bit address field. Initially this space was structured to give three different classes of unicast address with a different boundary between the identities of the host and the terminal port. The purpose of the classes was to manage the addressing space more efficiently given that there is a wide distribution of host network sizes (see Figure 4.14).

The fixed boundary was used for a period but when the rapid growth of the Internet started, putting special pressure on Class B addresses, the class definitions with their fixed boundaries were withdrawn in the early 1990s and replaced by the Classless Inter-Domain Routing (CIDR), allowing the boundary for each network to be adjusted to the requirements of the network plus a small allowance for growth. The /n that follows an address indicates the length in bits of the network identity.

The allocations of network identities were initially 'flat' and unstructured. This meant that routers needed to analyse the whole network part of the address in order to decide on routing. This resulted in the problem of the size of the routing tables in the routers growing faster than the capability of the router processors. To overcome this problem, CIDR also introduced the concept of aggregation at the higher level of the addresses. This meant that addresses were allocated so that all networks that were connected to the same backbone had the same early part (called prefix) of the address. This reduced the length of the number that needed to be analysed by most routers and reduced the number of routes that had to be announced to the outside world and included in routing tables.

At present with IPv4, aggregation has reduced the rate of growth of routing tables to a manageable rate (i.e. they are growing more slowly than processor capability).

Because aggregation was not started at the beginning and because the relationships between networks and backbone networks can change, there is not 100% aggregation in practice. The exceptions are called holes and should be avoided wherever possible as they require analysis of more bits in the address. However, the extent of the aggregation achieved is acceptable. Aggregation introduced the problem that the identity of the interconnected backbone network is contained in the identity of the network. If the interconnection arrangements change, either the addresses have to change or expensive holes are created in routing tables. Changes of addresses would require:

- changes in routing tables in the interconnected networks;
- changes in the resource records in the DNS system;
- changes in host addresses within the network.

Class	Format
A	7 bits for network identity; 24 bits for host identity
B	14 bits for network identity; 16 bits for host identity
C	21 bits for network identity; 8 bits for host identity

Figure 4.14 Classes in IPv4. © UMTS Forum (2001c).

The external changes to networks unavoidably require some effort to change routing tables and update the DNS system. Changes internally can be labour intensive for network administrators. A short-term solution for network administrators is provided by Network Address Translators (NAT) used at the boundary of the network, which enable a private addressing scheme to be used within the network. The private addressing scheme can remain unchanged when external connections change.

NATs are becoming quite widespread, and have been pushed heavily by vendors, but the IAB is against their use because they:

- 'Balkanise' the Internet, i.e. divide up what was meant to be a homogeneous whole into separate networks, thereby destroying the end-to-end model the Internet has been based on (however, it is uncertain whether some future market demands such as network integrity and billing will be compatible with the full end-to-end model);
- introduce a single point of failure (the NATs);
- fail to support applications that can use IP addresses at the application layer (actually bad practice).

A further method of reducing the demand for IP addresses is to use dynamic rather than permanent assignment of addresses. This solution is used by many dial-up ISPs, where addresses are allocated for each dial-in session. The protocols that support this are Dynamic Host Configuration Protocol (RFC 1541) and Point-to-Point Protocol (RFC 1661).

These developments, together with somewhat stricter assignment policies that now require evidence of need, have reduced the risk of the IPv4 address space being exhausted at least within the next 5 years with the current rate of growth.

4.5.4.2 Current Allocation Method for IPv4 Addresses

IP addresses are allocated by Regional Internet Registries (RIRs) in accordance with policies set by ICANN and its predecessor IANA. There are currently three RIRs although new ones, e.g. for Africa, may be created:

- Asia Pacific Network Information Centre (APNIC)
- American Registry for Internet Numbers (ARIN)
- Réseaux IP Européens (RIPE NCC), located in Amsterdam
- Latin American and Caribbean Internet Address Registry (LACNIC).

Each RIR allocates IP addresses to Local Internet Registries (IR), which are commonly ISPs. These Local IRs operate under the authority of the Regional IR and hold allocations for assignment to users. The term 'allocation' is used for space held by IRs for future assignment to users. Only assigned space is used by networks.

The goals of the allocation and assignment system are:

- uniqueness
- aggregation, to facilitate routing
- conservation
- registration.

Aggregation and conservation are sometimes conflicting because to maintain aggregation in a growth scenario spare space for growth must be included around allocations and if this space is not used eventually it may not be available for other users without creating a hole in the aggregation.

Two types of address space are used:

- provider aggregatable address space, where the aggregation is with respect to the connection to a backbone network;
- provider-independent address space, which may incur extra routing charges because of the additional complexity caused for routing tables and is assigned only under special conditions.

If a user with a provider-aggregated address space changes the interconnection arrangements, then he will have to release his address and obtain a new one. Any internal addresses that are visible to the public Internet will change.

4.5.4.3 New Arrangements for IPv6

IPv6 has a much larger address space than IPv4 and is structured differently so that aggregation is built in to the structure. Thus there is no provider-independent address space as there is with IPv4. The RIRs have published a draft set of common principles for IPv6 allocation. However, this draft received some criticism because its allocation strategy was too conservative and it is to be rewritten as soon as possible.

The allocation hierarchy of IPv6 is:

- ICANN
- RIRs, which allocate Top Level Aggregators (TLA) to Local Internet Registries (LIR)
- LIRs (also known as TLA Registries[5]), which may be transit operators and allocate Next Level Aggregators (NLA) and Site Level Aggregators (SLA)
- NLA registries, which are ISPs
- end-sites.

Because of the network hierarchy of IPv6, TLA registries, which are transit operators, carry out functions similar to those that RIPE NCC carried out for IPv4.

Figure 4.15 shows the main structure of IPv6 addresses in number of bits. Only 8192 TLA IDs are available for transit operators and so they need to be assigned with care. Consequently, a different initial structure will be used where one TLA value (0x0001) will be shared and sub-TLAs will be allocated out of it to the applicants for TLAs. These transit operators will use this allocation for assignments to ISPs (NLAs) until it is 80% used. Only then will they qualify for a full TLA assignment or a further sub-TLA. Figure 4.16 shows this initial structure as used by RIPE NCC.

For purposes of a 'slow start' of a sub-TLA, the first allocation to a TLA Registry may be a /35 block (representing 13 bits of NLA space). The Regional IR making the allocation will in this case reserve an additional 6 bits for the allocated sub-TLA. When the TLA Registry has

---

[5] TLA Registry means 'a registry that is a TLA', *not* 'a registry that allocates TLAs'.

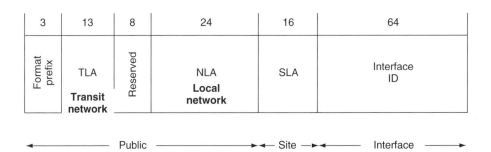

Figure 4.15 General structure of IPv6 addresses. © UMTS Forum (2001c).

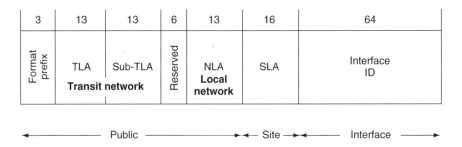

Figure 4.16 Initial structure of IPv6 address used by RIPE NCC under prefix (TLA 0x0001). © UMTS Forum (2001c).

fully used the first /35 block, the Regional IR will use the reserved space to make subsequent allocations to the same NLA.

Regional IRs will only make an initial allocation of sub-TLA address space to organisations that meet criterion (a) *and* at least one part of criterion (b), as follows:

a. The requesting organisation's IPv6 network must have exterior routing protocol peering relationships with the IPv6 networks of at least three other organisations that have a sub-TLA allocated to them.

AND either

b(i). The requesting organisation must have reassigned IPv6 addresses received from its upstream provider or providers to 40 SLA customer sites with routed networks connected by permanent or semi-permanent links.

OR

b(ii). The requesting organisation must demonstrate a clear intent to provide IPv6 service within 12 months after receiving allocated address space. This must be substantiated by such documents as an engineering plan or deployment plan.

For an initial bootstrap phase, b(i) is replaced by:

c. The requesting organisation must be an IPv4 transit provider and must show that it already has issued IPv4 address space to 40 customer sites that can meet the criteria for a /48 IPv6 assignment. In this case, the organisation must have an up-to-date routing policy registered in one of the databases of the Internet Routing Registry, which the Regional IR may verify by checking the routing table information on one of the public looking glass sites.

OR

d. The requesting organisation must demonstrate that it has experience with IPv6 through active participation in a trial network for at least six months, during which time it operated a pseudo-TLA (pTLA) for at least three months. The Regional IRs may require documentation of acceptable routing policies and practice from the requesting organisation.

TLA registries must register all end-sites.

4.5.4.4 IPv4 and IPv6 Compatibility

IPv4 uses 32-bit addresses whereas IPv6 uses 128 bits. The two forms of address are therefore only compatible in one direction (RFC 1933, 'Transition mechanisms for Ipv6 hosts and routers').

Networks are built using either one or both technologies. DNS will provide a separate record for each type of IP address. It will provide:

- an 'A' record, giving an IPv4 address, or
- an 'AAAA' record giving an IPv6 address

depending on the information provided to DNS by the ISP that is serving the name in question.

Two compatibility issues arise:

- What incoming traffic can be terminated?
- How can traffic be sent across networks of a different type?

Compatibility
The compatibility relationships are as shown in Figure 4.17.

Incompatibility happens when a user on an IPv4 domain wishes to communicate to an IPv6-only domain, because the IPv4 domain cannot handle the IPv6 address that is to be sent. The only way to solve this problem is if the IPv6 address is made compatible with IPv4 by using only the first 32 bits in the header and setting the remaining 96 bits to zero. This is not an adequate solution because the objective of IPv6 is to provide more addresses than are available within IPv4.

A network that uses only IPv6 will be able to send packets to networks that implement IPv4 because it can put the IPv4 address in the front part of the IPv6 space in the packet but the receiving network will need to know how to handle the incoming packet.

	Receiving network technology		
Sending network technology	IPv4	IPv4 + IPv6	IPv6
IPv4	Yes	Yes	No because IPv4 cannot handle IPv6 address
IPv4 + IPv6	Yes	Yes	Yes
IPv6	Possibly because IPv4 address can be handled within IPv6 address	Yes	Yes

Figure 4.17   Incompatibility table. © UMTS Forum (2001c).

IPv6 across IPv4-only Networks

IPv6 traffic can be carried across IPv4-only networks by encapsulating IPv6 packets by adding an IPv4 address to the front where the IPv4 address indicates the destination router or host. This router or host will then strip off the additional IPv4 address. This technique creates a tunnel across an IPv4 only domain and is called configured tunnelling.

In order to use configured tunnelling, routers at the border of IPv6 domains will have to be assigned to IPv4 addresses that can be used as the end-points of tunnels.

Various transition tools are described in RFC 1933.

4.6 IMS

IMS has been described from the service point of view in Chapter 3. Here, the technological background is described.

4.6.1 TECHNOLOGY ASPECTS

Some other advantages that IMS brings are revealed when you examine the technology. To answer the question of when IMS should be deployed, you need to understand what is involved in implementing IMS. You need to understand the essential technology enablers that are required. Then you will need to estimate the cost of implementing IMS, as well as potential savings, and then compare this to the revenues generated by the subscription and billing models for IMS-enabled services – in fact, conduct a business case and finally determine a deployment strategy for the industry.

To start off our understanding of IMS and what is involved in deploying it, we will look at the basics of the IP multimedia subsystem (IMS) and also Voice over IP (VoIP).

The technology aspects that are fundamental to IMS and VoIP are:

- separation of control and bearer planes;
- IP multimedia call control;
- quality of service (QoS).

4.6.2 SEPARATION OF CONTROL AND BEARER PLANES

In a conventional telephone or mobile network, the control or signalling shown by the dotted line in Figure 4.18 follows exactly the same physical path through the network as the user traffic, or bearer, shown by the full line.

The process of controlling the call – for example, set-up or clear down – and the process of selecting the route is performed by the same platforms in the network – the switches.

In a VoIP or IMS network, which is basically a packet network rather than a circuit switched network, the traffic is sent as packets whose destination is determined physically by routers. The process of controlling the call is performed by a call server, which is physically separate from the routers, but provides them with instructions on where to send packets. The control and bearer or user path are now separate (Figure 4.19).

In fact, the call server does not even have to be in the same network (Figure 4.20). It could be anywhere on the connected Internet or even in a different country. The call control is based on Internet protocols and can process multimedia calls. 3GPP chose call control for IMS based on the session initiation protocol (SIP) specified by IETF.

4.6.3 IP MULTIMEDIA CALL CONTROL

SIP is standard IP server software written in standard software language. Toolkits and development products will be available to enable new services to be easily developed – they

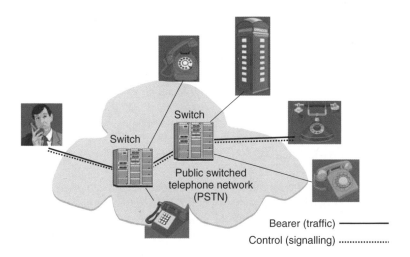

Figure 4.18   Separation of control and bearer – conventional. © UMTS Forum, 2002.

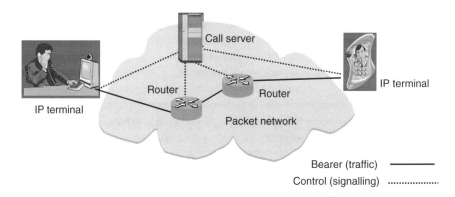

Figure 4.19   Separation of control and bearer – VoIP or IMS (packet) network. © UMTS Forum, 2002.

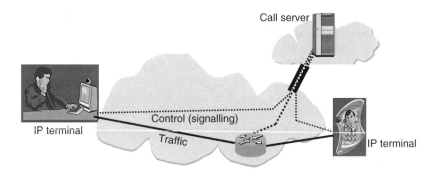

Figure 4.20   IP multimedia call control. © UMTS Forum, 2002.

will be platform manufacturer and network independent. Large teams of software writers producing code dedicated to a specific ISDN switch platform are no longer needed. Servers and codes for individual services and applications are largely independent and do not interact, which means that the huge testing programmes that are necessary with switches can be avoided. Service software does not have to be installed in every switch platform in the network – just one or a few servers, which can be placed anywhere in the network or outside it. This is termed 'edge of network control'. This all means that innovative telephony services can be created quickly, easily and cheaply by entrepreneurs – just like Internet services.

In IMS and VoIP speech telephony is moved from the circuit-switched domain to the packet-switched domain (Figure 4.21).

This means that there is one common network for speech, data and information services to manage, which can save on operating costs. Platforms are based on industry standard routers and servers. That means that they should be able to lower cost. To interwork with conventional networks and terminals, media gateways and soft-switches (switch servers) are introduced to process the traffic and signalling, respectively.

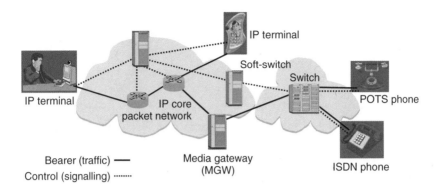

Figure 4.21 Speech telephony moving from circuit-switched domain to packet-switched domain. © UMTS Forum, 2002.

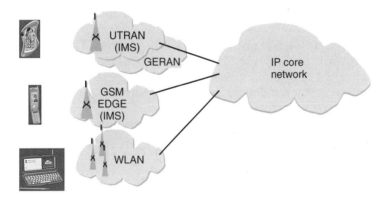

Figure 4.22 Access independance. © UMTS Forum, 2002.

IP networks will deal with, for example (Figure 4.22):

- UMTS networks based on UTRAN;
- American 3G networks based on the GERAN;
- classical GSM or new UMTS networks based on EDGE technology;
- wireless LAN.

At last these networks can all be based on a common core network.

4.6.4 IMS: QUALITY OF SERVICES

Microsoft NetMeeting, the well-known IP multimedia network application, serves as an example of QoS for real-time speech and multimedia. The current versions of NetMeeting are based on an earlier IP multimedia call control, H.323 from ITU rather than SIP from IETF.

However, it is moving to SIP as the two standards converge and SIP is now included in the Microsoft XP operating system. NetMeeting enables users in a live Internet conference to share:

- presentations, like PowerPoint;
- text documents, like Microsoft Word;
- text 'chat', either one-to-one or one-to-group.

The graphics shown on all the remotely stationed group members' displays can be controlled by the chairperson. It also has a built-in conference phone facility, using the Internet connection. It could therefore be considered the forerunner of IP multimedia applications. Without QoS management in the network most Internet connections are good enough for the graphics and text – the users' screens appear sufficiently well synchronised. In some cases, it is good enough for the speech part – for example, in a lightly loaded intranet with a LAN connection. However, when the Internet connection is slow or becomes heavily loaded there is a problem. If someone starts a large file download, the network becomes congested, for example at the router shown. The Internet or 3G network does not know the difference between critical speech packets and the less critical graphics, text and data traffic. Without QoS management the network can not prioritise packets and the congestion causes the packets to be delayed equally, and the speech packets in particular suffer delay and jitter. The speech breaks up and becomes unintelligible. So today, most NetMeeting sessions use the Internet connection for the graphics and text – delay and jitter have little noticeable effect on these. And they use a completely separate connection through the telephone network for the speech conversation, using an ordinary conference phone.

However, with IMS it would be one simple integrated set-up process. When QoS is managed in VoIP and IMS networks, QoS works to ensure that critical packets such as speech are prioritised over less critical data packets (Figure 4.23).

QoS requirements for speech packets are demanding:

- Delay should be less than 150 ms with an absolute maximum of 400 ms on international circuits.
- Jitter must be less than 1 ms.
- Packet loss rate must be less than 0.1%.

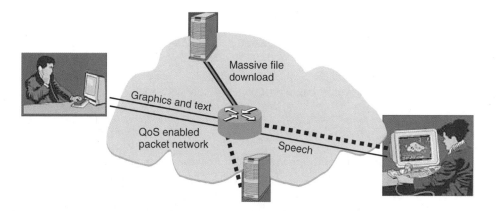

Figure 4.23   Managed QoS in VoIP and IMS networks. © UMTS Forum, 2002.

There are basically two key enablers to realise QoS. First, Multi-Protocol Label Switching (MPLS) is an enhancement to the routers which simplifies and speeds up the routing function locally and provides an interface to allow routing to be managed at the network level.

Second, Resource Reservation Protocol (RSVP) is call control software which works in conjunction with the router interface to enable paths to be defined that avoid congestion and meet the specific requirements of the media being transported.

VoIP technology is also important to mobile networks. MPLS and RSVP are techniques that are inherent in IMS. The mobile operator may need to upgrade its own fixed Internet backbone with MPLS and RSVP in order to support IMS. Inter-network links must be QoS enabled in order to provide the necessary subscriber-to-subscriber voice quality. This is an important roll-out consideration.

Roll-out across different networks needs to be considered (Figure 4.24). In a world of connected networks, you suppose that only mobile network A upgrades to 3G. The subscriber in network A can have the improved 3G information services advertised, whether or not any other networks upgrade to 3G, and the subscriber is satisfied. Media gateways allow the subscriber to communicate with users in the other networks, but he cannot have the promised rich new IMS features in calls he has with them. The subscriber is likely to be dissatisfied.

Because most calls in mobile networks connect with subscribers in outside networks, it is important that the introduction of VoIP and IMS is harmonised across all networks to ensure that there are plenty of users with IMS features that the user can call.

Consider now the special technology enabling requirements for mobile networks and the 3GPP IMS network architecture. There are additional considerations for mobile IP networks: security, privacy, authentication, registration, billing, roaming, handover, location, legal intercept etc. All these requirements have shaped the 3GPP IMS network architecture.

4.6.5 UMTS BASELINE NETWORK ARCHITECTURE

UMTS/3G is the baseline against which IMS deployments will be compared. It happens also to correspond to one of the deployment scenarios – 'no IMS'. A generic network architecture is shown in Figure 4.25. It is based on Release 99 of the UMTS standard published by 3GPP. No attempt has been made to model a real physical deployment or to scale the relative numbers of the different network elements required. The platforms needed to provide services and

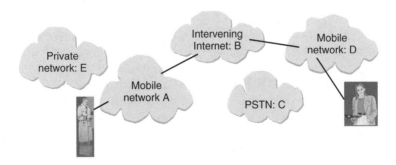

Figure 4.24 Roll-out considerations. © UMTS Forum, 2002.

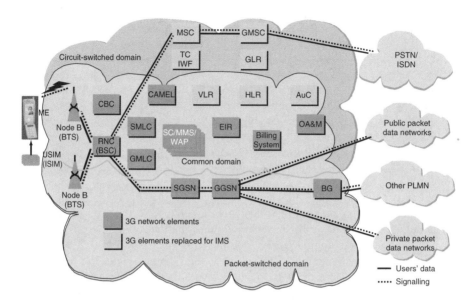

Figure 4.25   3GPP Release 99 UMTS/3G reference network architecture. © UMTS Forum, 2002.

content are also shown, as well as the end-users' traffic flow. However, the complex signalling data flow is only shown by the network 'clouds', apart from an initial connection between the mobile network (PLMN) and each of the external networks. It shows a fundamental feature of pre-IMS networks, namely that the signalling data follow the same path as the traffic. The call control and service provisioning software are both integrated in the same platforms, making the development of new telephony services costly and time consuming.

The architecture presented now for the main comparison with 3G assumes that the existing 3G/UMTS network is upgraded to include IMS, including multimedia call control based on SIP from IETF, and the legacy mobiles supporting circuit-switched voice are still required. That corresponds to 'full IMS'.

A complete generic network architecture is shown in Figure 4.26. It is based on 3GPP Release 5 of the UMTS standard. It shows the platforms that have been added in order to implement the IMS core network as well as the 3G platforms that require substantial upgrade in order to support it. These are mainly platforms within the packet-switched domain.

IMS does not necessarily put additional requirements on the components of the infrastructure that transport the IP packets. An installed base using ATM transport ('IP over ATM') is capable of supporting IMS as an alternative to one based on all-IP transport. A feasible path for an operator is to introduce all-IP transport when it is justified for other reasons in addition to IMS. For example, network expansion can be done with IP transport while keeping the installed ATM transport. Thus the overall business plan can decide when the ATM transport network should be replaced by all-IP. However provided, the backbone packet transport network must support the QoS requirements for IMS and upgrades may be necessary in some cases.

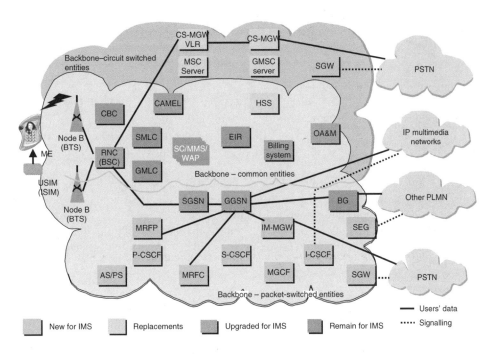

Figure 4.26 3GPP Release 5 UMTS/3G reference network architecture – with IMS. © UMTS Forum, 2002.

A 3G network comprises a circuit-switched domain, a packet-switched domain and a domain common to both packet- and circuit-switched. There follows an explanation of the network elements as shown in Figures 4.25 and 4.26.

First, the network elements of the circuit-switched domain and network elements common to circuit and packet:

- MSC: the Mobile Switching Centre is the main interface between the radio system and the fixed telephone network. Its fundamental purpose, like a fixed network switch, is to switch the connections between subscribers, but it differs in that it has to perform location registration and handover procedures in order to serve the mobility of the subscribers.
- GMSC: the Gateway MSC has capability, in addition to that of an MSC, to select and interrogate HLRs in order to route a call from the PSTN to the correct MSC for the wanted mobile subscriber.
- TC/IWF: the Transcoder and Interworking Function performs conversions between the spectrally efficient, robust speech coding (TC) and formats for circuit-switched data services (IWF), used over the radio interface and those used in the fixed networks.
- HLR: the Home Location Register is a database containing key information about subscribers who belong to the network. It includes subscription information (i.e. which services the user is entitled to use) as well as some information on the subscriber's current location, enabling incoming calls to be correctly routed to him/her and

charging information to be sent to networks that the subscriber roams to. The HLR is one of the key components that enable mobility and roaming in GSM, GPRS and UMTS networks.

- GLR: the Gateway Location Register (optional) handles location management of the roaming subscriber in a visited network without involving an HLR.
- VLR: the Visitor Location Register is a dynamic database that contains information about mobiles which have roamed into its area, in particular, the Location Area in which the mobile has registered itself. The VLR in a visited network exchanges information with the HLR in the subscriber's home network to enable the correct routing and handling of calls.
- AuC: the Authentication Centre is a database that is associated with each HLR. For every subscriber registered with the HLR, it stores an identity key and uses it in the processes of authenticating the mobile and ciphering its transmission.
- EIR: the Equipment Identity Register stores the International Mobile Equipment Identity (IMEI) number for each mobile, classifying it as either 'white listed', 'black listed' or 'grey listed'.

Second, the network elements of the Radio Access Network (RAN):

- RNC or BSC: the Radio Network Controller (RNC) provides the control for a network of UMTS base stations that are based on WCDMA technology. The Base Station Controller (BSC) provides a similar function in EDGE-based UMTS networks. The principal function provided is the real-time handover of mobiles between base stations.
- BTS or Node B: the Base Transceiver Station (BTS) provides the radio transmission and reception for one cell in EDGE-based UMTS networks. In UMTS networks based on WCDMA, the base station element serving one cell is called a Node B.
- MS: the Mobile Station comprises two main parts, the mobile equipment and the UMTS Subscriber Identity Module.
- ME: the Mobile Equipment is the main physical, hardware and software part of a mobile terminal.
- USIM: the UMTS Subscriber Identity Module is usually a removable device, which stores the subscriber's profile (e.g. his/her service provider, phone number, subscription data, as well as personal data such as directories and stored short messages). The USIM performs important functions in connection with the authentication of subscribers.

Third, the network elements of the packet-switched domain:

- SGSN: the Serving GPRS Support Node, together with the GGSN, constitutes the interface between the radio system and fixed packet data networks (including the Internet and intranets). The SGSN stores routing information to the cell or area where the mobile is located plus its associated VLR.
- GGSN: the Gateway GPRS Support Node communicates with the HLR and stores routing information to the SGSN where the mobile is registered.
- BG: the Border Gateway is an interface to the external inter-PLMN links connecting to other GPRS (or UMTS PS) networks. It provides security to protect the PLMN and its subscribers.

4.7 WLAN/Wi-Fi

Ever since the Ethernet Project emerged from the Xerox Palo Alto Research Center in the early 1970s and other similar digital protocols quickly followed, the basic technology has been in place for local area networks (LANs) to blossom in both the public and private sectors. Standard LAN protocols, such as Ethernet, which operate at fairly high speeds with inexpensive connection hardware, can bring digital networking to almost any computer. Today, organisations of every size access and share information over a digital network; the power of networking and collaborative, distributed computing is beginning to be realised. However, until recently, LANs were limited to the physical, hard-wired infrastructure of the building. Even with dial-up and leased lines, network nodes were limited to access through wired, landline connections. Typical network users e.g. mobile users in businesses and universities, to name but a few, would benefit greatly from the added flexibility and mobility capabilities of wireless LANs (Lough *et al.*, n.d.).

In the mid-1990s, IEEE started to develop a wireless access technology for short-range radio delivering high speed data rates to connect computers directly to the LAN without cable connection. This technology should make offices, business parks and other business areas more flexible in connecting new computers to their network without pulling new cables through the building. After a slow start to the standardisation process, IEEE's working group P802.11 agreed on three physical layer (PHY) solutions: two radio-based – one with frequency hopping (FH) and one with direct sequence (DS) – and one infrared, using respectively the ISM[6] frequency band at 2.4 GHz and the IR band at 300–428 GHz . This IEEE 802.11 standard was completed in 1997. Output power was limited to 100 mW, which should allow data rates of 200 Kbps up to 1 Mbps with operating frequencies coexisting with industrial, scientific and medical applications in the unlicensed frequency bands in the 2.4 and 5 GHz.

Soon derivates were developed, all with different goals and performance, but some aspects of the network architecture need to be understood. There is, first, the IEEE 802.11 infrastructure. Generally, any IEEE 802.11-equipped laptops communicated with each other and exchange data, e.g. via IR point-to-point. The same would apply to Bluetooth/UHF-equipped PCs. If there are three or more PCs which require to interconnect randomly (or 'on the fly') then more organisation is needed, e.g. algorithms such as the spokesman election algorithm (SEA). These have been designed to 'elect' one machine as the base station (master) of the network with the others being slaves.

More important and useful are WLAN networks, working in the same way as an Ethernet. Ethernet architecture uses fixed network access points with which mobile nodes can communicate. These network access points are sometime connected to landlines to widen the LAN's capability by bridging wireless nodes to other wired nodes. If service areas overlap, handoffs can occur. This structure is very similar to the present day cellular networks around the world. But to work in WLAN it needs some additional agreements to be set up in the Medium Access Control (MAC) layer of the network.

As described in by Lough *et al.* (n.d.), the MAC layer is a set of protocols which is responsible for maintaining order in the use of a shared medium. The 802.11 standard

---

[6] ISM stands for 'industrial, scientific and medical' use of the dedicated frequency band.

specifies a carrier sense multiple access with collision avoidance (CSMA/CA) protocol. In this protocol, when a node receives a packet to be transmitted, it first listens to ensure no other node is transmitting. If the channel is clear, it then transmits the packet. Otherwise, it chooses a random 'backoff factor', which determines the amount of time the node must wait until it is allowed to transmit its packet. During periods in which the channel is clear, the transmitting node decrements its backoff counter. (When the channel is busy it does not decrement its backoff counter.) When the backoff counter reaches zero, the node transmits the packet. Since the probability that two nodes will choose the same backoff factor is small, collisions between packets are minimized. Collision detection, as is employed in Ethernet, cannot be used for the radio frequency transmissions of IEEE 802.11. The reason for this is that when a node is transmitting it cannot hear any other node in the system which may be transmitting, since its own signal will drown out any others arriving at the node.

Whenever a packet is to be transmitted, the transmitting node first sends out a short ready-to-send (RTS) packet containing information on the length of the packet. If the receiving node hears the RTS, it responds with a short clear-to-send (CTS) packet. After this exchange, the transmitting node sends its packet. When the packet is received successfully, as determined by a cyclic redundancy check (CRC), the receiving node transmits an acknow-ledgment (ACK) packet. This back-and-forth exchange is necessary to avoid the 'hidden node' problem. This describes a condition where between three nodes A, B, C, node A, for example, can communicate with node B, and node B can communicate with node C. However, node A cannot communicate with node C. Thus, for instance, although node A may sense the channel to be clear, node C may in fact be transmitting to node B. The protocol described above alerts node A that node B is busy, and hence it must wait before transmitting its packet.

The other members of the 802.11 family should also be discussed. The first derivate of importance is IEEE 802.11b, which is an evolution of the basic IEEE 802.11. It uses the same frequency band, the same number of channels, one of the three modulation schemes (DSSS=Direct Sequence Spread Spectrum), which makes it compatible with IEEE 802.11, but e.g. it uses 5-MHz carrier channel spacing, which makes it interoperable in CEPT countries. There are some other modern features (size, power consumption etc.) which make it an improved evolution of IEEE 802.11.

In parallel with the development of IEEE 802.11b, the derivate IEEE 802.11a was developed. IEEE 802.11a uses a different physical layer with orthogonal frequency division multiple access modulation (OFDMA). With that scheme it achieves 54-Mbps bit rates and operates in the 5 GHz band on eight channels.

A synthesis of both is IEEE 802.11g, which is in a sense an evolution of IEEE 802.11a and b. Using the same physical layer as in standards IEEE 802.11a and b, depending on bit rates, it is in many parameters compatible to IEEE 802.11a and b. To combine the advantages of both standards IEEE 802.11a and b, the new IEEE 802.11g is going to be widely implemented in new PC, laptop and PDA equipment and is likely to increase further the 802.11 WLAN market share.

As the IEEE 802.11 infrastructure and systems technology has not yet reached the same level as cellular mobile systems, IEEE is seeking to further extend performance and features. Examples are:

- IEEE 802.11e provides improved QoS with better multimedia access and VoIP in connection with cellular phones.
- IEEE 802.11f provides handover procedures which will make WLAN products able to move from one cell to another without losing the link.
- IEEE 802.11i provides a new security algorithm, called Advanced Encryption Standard (AES).

Important improvements for WLAN security and performance were reached in 2004. The 802.11i standard is the finished version of the Wi-Fi Protected Access (WPA) security standard with the added security of AES. The IEEE ratified the 802.11i security protocol in June 2004 (Schwartz, 2004). 802.11i is a highly anticipated update to 802.11 security, which includes all of the measures found in Wi-Fi Protected Access (WPA), as well as a longer, stronger encryption key (128-bit cipher). The new key, though, could mean that some older Wi-Fi equipment will need to be replaced in order to work with 802.11i. The Wi-Fi Alliance said it will certify 802.11i under the name WPA2. The Wi-Fi Alliance started certifying 802.11i and 802.11e products by late 2004 (IDG News Service and CTIA Daily News, 2004). The Wi-Fi Alliance will also certify products using Wireless Media Extensions (WME) technology, a subset of 802.11e. The 802.11e standard improves wireless network quality for use with voice and data services. It does so by giving precedence to transmissions that must go through without interruption, such as streaming video and voice.

This summary of WLAN would not be complete without mentioning some European Standards on WLAN (Huber and Huber, 2002). The European Research Programme BRAIN (Urban *et al.*, 2001) of the European Telecommunication Standardisation Institute (ETSI) developed the so-called HIPERLAN standards (High-Performance Local Radio Access Network).

HIPERLAN/1 provides bit rates of up to 23.5 Mbps, uses the 5 GHz band provided by the European Radio Committee (ERC), designated for use on a regional and national basis, and is compatible with Ethernet and Token Ring LAN standards according to ISO 8802.3 and 8802.5. The output power is limited to 200 mW (5.15–5.35 GHz) or 1 W (5.47–5.725 GHz), depending on frequency. User mobility is restricted within the local service area only.

HIPERLAN/2 provides up to 25 Mbps and is foreseen to be linked with IP networks. Typically, it will work only indoors and mobility lies within the local area. The modulation scheme is different from HIPERLAN/1, using OFDMA. Its advantage over the IEEE standards lies in better security and individual authentication.

The technical specification was released by ETSI in 1997. In contrast to its American rival, it is more complete and complicated. It offers acceptable security, handover and roaming procedures right now. It is hard to say whether it can enter the market successfully now that IEEE 802.11x has already become a de facto world standard for WLAN.

With these remarks we conclude the summary of technical matters that are relevant to the business chapters of this book.

5

Spectrum for UMTS

Before we go on to discuss spectrum for UMTS, a few remarks about the fundamentals of spectrum use are in order. As oil is to the automotive industry, so spectrum of frequencies is to the telecommunication industry. Both are scarce resources and the competition to get the best slice of the cake is relentless. For each radio service you need a specific spectrum of frequencies, and as radio waves do not respect international borders worldwide frequency coordination is a very important task. As mentioned in Chapter 6, the international community has agreed to handle this through the ITU. Under their rules, member states distribute radio frequencies through national licences. Countries have agreed to accept and respect the frequency plan given and coordinated by ITU. So spectrum and radio service coordination is an ongoing issue for all regulatory bodies. An example of frequency bands for mobile communications is shown in Figure 5.1.

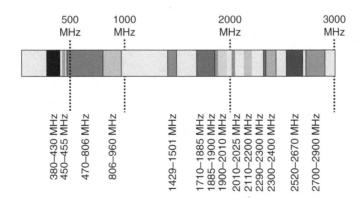

Figure 5.1 Mobile communication frequencies in the 380–3000 MHz band.

The Mobile Multimedia Business: Requirements and Solutions Bernd Eylert
© 2005 John Wiley & Sons, Ltd

Before continuing, the reader will be given a brief introduction to the relevant organisations. First, ITU and WRC/WARC: the ITU, headquartered in Geneva, Switzerland is an international organisation within the United Nations where governments and the private sector coordinate global telecom networks and services. The three sectors of the ITU – Radiocommunication (ITU-R), Telecommunication Standardization (ITU-T), and Telecommunication Development (ITU-D) – work today to build and shape tomorrow's networks and services. Their activities cover all aspects of telecommunication, from setting standards that facilitate seamless inter-working of equipment and systems on a global basis, to adopting operational procedures for the vast and growing array of wireless services and designing programmes to improve telecommunication infrastructure in the developing world (www.itu.int/aboutitu/overview/role-work.html). The ITU runs frequently the World Radiocommunication Conferences (WRC), until 1992 the World Administration Radiocommunication Conference (WARC), to coordinate frequency plans and spectrum worldwide. ITU-T is the successor of the previous Comité Consultatif International de Téléphone et Télégraphique (CCITT), while ITU-R followed Comité Consultatif International de Radiocommunication (CCIR).

Second, CEPT, ETSI, ETNO, CERP, ERC, ECTRA, ERO and ECC: the European Conference of Postal and Telecommunications Administrations (CEPT) was established in 1959 by 19 countries, which expanded to 26 during its first 10 years (www.cept.org). Original members were the incumbent monopoly-holding postal and telecommunications administrations. CEPT's activities included cooperation on commercial, operational, regulatory and technical standardisation issues.

In 1988 CEPT decided to create the European Telecommunications Standards Institute (ETSI), to which all its telecommunication standardisation activities were transferred.

In 1992 the postal and telecommunications operators created their own organisations, Post Europe and ETNO, respectively. In conjunction with the European policy of separating postal and telecommunications operations from policy making and regulatory functions, CEPT thus became a body of policy makers and regulators. At the same time, Central and Eastern European countries became eligible for membership of CEPT. With its 45 members CEPT now covers almost the entire geographical area of Europe.

CEPT after this agreement deals exclusively with sovereign/regulatory matters, and has established three committees, one on postal matters, Comité Européen de Réglementation Postale (CERP), and two on telecommunications issues: the European Radiocommunications Committee (ERC) and the European Committee for Regulatory Telecommunications Affairs (ECTRA).

The European Radiocommunications Office (ERO) was formally opened on 6 May 1991 and is located in Copenhagen, Denmark. ERO is the permanent office supporting the Electronic Communications Committee (ECC) of CEPT. ERO was established on the basis of a Memorandum of Understanding (MoU). In 1996 this MoU was replaced by the 'Convention for the establishment of the European Radiocommunications Office', which has to date been signed by 30 CEPT administrations and which defines the terms of reference for ERO and the funding arrangement (www.ero.dk).

Some of these frequency bands shown in Figure 5.1 will play a significant role in the following discussion. First, the spectrum for Future Public Land Mobile Telecommunication System (FPLMTS), the concept for a new mobile communication system available for the new millennium (2000), is introduced as it was allocated in WARC-92 by ITU. A discussion on additional spectrum for UMTS will follow, based on market demand, sometimes called Extension Bands, which was extensively prepared especially by the UMTS Forum and ERO for WRC-2000 and allocated as IMT-2000 by ITU. With this in mind the reuse of frequencies for new services, some scenarios for the use of the new additional band for IMT-2000 and further plans for WRC-07 and more spectrum for 3G and beyond will be discussed. In this chapter, the traffic characteristics as studied by the Forum in the preparation of WRC-2000 will be explained, but we will also have a look at the latest views which the industry has on this subject, with a focus on 2010 and beyond.

It should be noted that the discussions and results of this chapter are inter alia based on UMTS Forum reports nos 1, 5, 6, 7 and 33 on this subject (UMTS Forum, 1997, 1998c,d, 1999a, 2003b).

5.1 SPECTRUM AVAILABILITY

5.1.1 WARC-92 INITIAL ALLOCATION

The 1992 WARC Conference (WARC-92) in footnote no. S5.388 of the Radio Regulations,[1] identified frequency bands 1885–2025 and 2110–2200 MHz, i.e. a total of 230 MHz, for IMT-2000 (previously known as FPLMTS). This allocation was based on calculations done before WARC-92 which took into account speech and low data rate services as a major source of the traffic in Recommendation ITU-R M.687-2. Several more recent investigations (UMTS Forum, 1998d; ITU-R, 1998a,b) have shown that due to the tremendous growth of mobile communications and multimedia since 1992, the initial WARC-92 allocation would support only the initial introduction of UMTS/IMT-2000 systems. Because of the growth in both the number of users and the bandwidth they require, this initial spectrum will be insufficient to support further development.

The following ITU-R documents give more information concerning the ITU-R situation related to IMT-2000 spectrum aspects:

- ITU-R Resolution 212 originated by WARC-92, revised by WRC-97, summarises the developmental, spectrum and regulatory issues pertinent to IMT-2000.
- ITU-R Recommendation M.816 considers the framework for a wide range of services supported by IMT-2000.
- ITU-R Recommendation M describes a detailed methodology for the calculation of terrestrial spectrum requirements taking into account all new wideband services.
- ITU-R Report M gives the total spectrum requirement for terrestrial mobile services in year 2010 applying the above mentioned methodology.

---

[1] The bands 1885–2025 MHz and 2110–2200 MHz are intended for use, on a worldwide basis, by administrations wishing to implement the future public land mobile telecommunication systems (FPLMTS). Such use does not preclude the use of these bands by other services to which the bands are allocated. The bands should be made available to FPLMTS in accordance with resolution 212 (REv WRC-95).

WRC-97 recognised the additional spectrum needs for IMT-2000 by including this issue as Agenda Item 1.6 of the next WRC:

- 1.6.1 – review of spectrum and regulatory issues for advanced mobile applications in the context of IMT-2000, noting that there is an urgent need to provide more spectrum for the terrestrial component of such applications, and priority should be given to terrestrial mobile spectrum needs, and adjustments to the table of frequency allocations as necessary.
- 1.6.2 – identification of a global control channel to facilitate multimode terminal operation and worldwide roaming of IMT-2000.

The spectrum, initially made available for operators, is for most of the worldwide regions and areas the same as that which was identified for FPLMTS/IMT-2000 at WARC-92 and in the ITU Radio Regulations. Figure 5.2 shows the IMT-2000 spectrum situation in some countries and areas before WRC-2000.

As shown in Figure 5.2 some countries have not converged fully in the direction of the WARC-92 agreement.

5.1.2 EUROPEAN SPECTRUM FOR UMTS

The European spectrum designation is defined in the CEPT ERC Decision on the introduction of UMTS (CEPT ERC/DEC/(97)07).

CEPT/ERC/TG1 produced this ERC Decision on the harmonised utilisation of spectrum for terrestrial Universal Mobile Telecommunications System (UMTS) operating within the

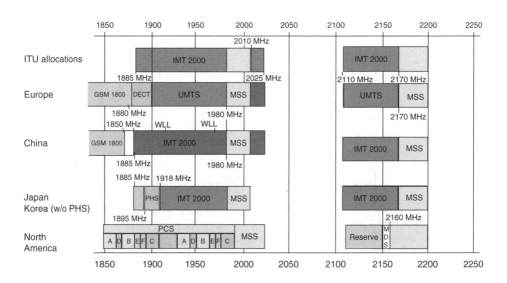

Figure 5.2 Frequency bands for FPLMTS/IMT-2000 agreed at WARC-92. © UMTS Forum (1998c).

bands 1900–1980 MHz, 2010–2025 MHz and 2110–2170 MHz (ERC, 1998). The purpose is to facilitate efficiency in utilisation of the UMTS bands across CEPT by:

- identifying a common approach to spectrum planning;
- encompassing spectrum allocated on an exclusive basis for public licensed UMTS networks;
- encompassing spectrum identified for shared use by license-exempt applications.

In its regulatory capacity the ERC also decided that regulatory authorities in Europe should guarantee that at least 2×40 MHz of the allocated UMTS frequency band should be available to operators in 2002. As a consequence different operator scenarios developed in this report may help to deal with such situations. Finally, to pick up the result in advance, regulators in Europe reached that goal on time.

CEPT in Europe can make available all ITU spectrum except 15 MHz which are already used for DECT. This results in 155 MHz of spectrum for terrestrial services with an additional 60 MHz identified for UMTS satellite services within the 2 GHz mobile satellite service (MSS) bands, available up to 2005 subject to market demand.

ERC/TG1 has led the Conference preparations for IMT-2000 in CEPT. ERC/TG1 calculated the spectrum requirement for UMTS/IMT-2000 and it has identified the European candidate bands for the extension bands. All the candidates are in the Conference Preparatory Meeting (CPM) list; the most valuable are listed in Section 5.3.5.

5.1.3 UMTS SPECTRUM IN OTHER AREAS OF THE WORLD

5.1.3.1 Asia-Pacific

The spectrum allocation in the Asian Pacific states is very similar to that in Europe. In 1999, the Japanese Ministry of Post and Telecommunications, MPT, designated the WARC-92 spectrum for third-generation systems in the same way as the Europeans with the difference that the frequency band 1895–1918.1 MHz is already allocated to Personal Handy-Phone System (PHS) services. Three 3G licences were granted in 2000, shortly after WRC-2000, each 2×20 MHz, which covers the whole paired band as allocated by WARC-92. Korea indicated spectrum allocations for IMT-2000 in line with spectrum identified for IMT-2000 in the ITU Radio Regulations and granted three licences in August 2001. Most of the so-called 'tiger' countries followed the Japanese and Korean format in 2001 and made paired and unpaired bands available. Therefore, similar operator scenarios will appear as in Europe. In China the major part of the ITU bands could be made available, and it is believed it will be in place for licences to be granted by the middle of this decade.

5.1.3.2 North America

There is a different situation in North America. The introduction of personal communication services (PCS) and their auctioning led to a split into spectrum licences of 2×15 MHz and 2×5 MHz in the band from 1850 MHz up to 1990 MHz (Figure 5.2) by the mid-1990s. This differing spectrum utilisation leads to questions about how IMT-2000 services can be implemented and how radio equipment could be harmonised with IMT-2000 services in Europe,

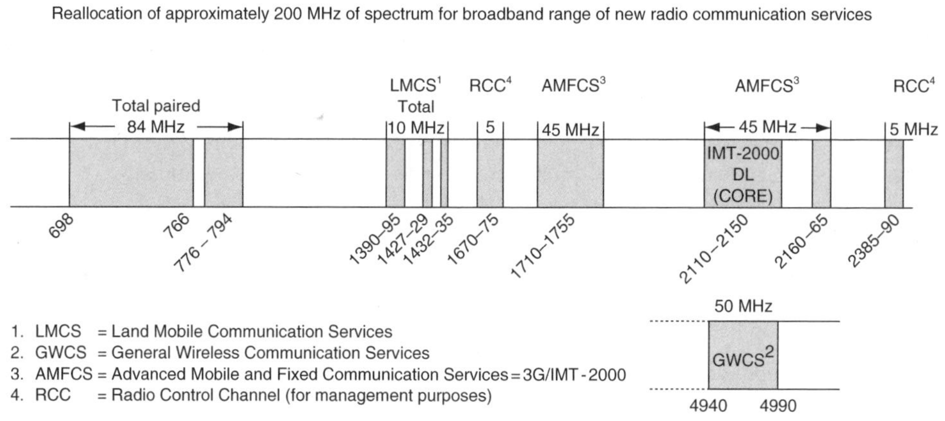

Reallocation of approximately 200 MHz of spectrum for broadband range of new radio communication services

Figure 5.3   FCC policy statement for spectrum management, 18 November 1999. *Source*: FCC News.

Japan and the rest of the world. As shown in Figure 5.2, North America can use only 5 to 15 MHz frequency blocks in the PCS bands. Therefore the 5 MHz minimum bandwidth per operator is an important requirement for the standard; it also has to include guard band.[2] The upper part of the IMT-2000 spectrum from 2110 to 2160 MHz could potentially be used by IMT-2000/UMTS (Figure 5.3).

With the so called 'Clinton Memorandum' of September 2000 the world had hoped for worldwide harmonisation of the IMT-2000 frequency bands. President Clinton had called on the American regulators to adopt ITU recommendations regarding the frequency bands, thus the 1900 MHz frequency band should have been available for UMTS networks in the United States from 2001 onwards with a general start of services in 2004. But with the presidential election in November the same year, US spectrum policy turned away from these reforms. In any event, the FCC found a new home – the 2.5 GHz band – for 2.1 GHz users forced off the band to free spectrum for third-generation wireless systems (Silva, 2004).

Canada has kept the C and E frequency blocks as a reserve for later allocations.

5.1.3.3 Latin America

Latin America includes Central America and the Caribbean and represents 35 countries in total, of which 16 have allocated the 1900 MHz band for PCS. CITEL, the Inter-American Telecommunications Commission, was seeking global spectrum for IMT-2000 by 2005 through its PCC-III process. The USA and Mexico decided to use the lower GSM 1800 band (1710–1785 MHz) paired with the upper IMT-2000 band (2110–2170 MHz) and, of course, the original WARC-92 band, dedicated for FPLMTS, for 3G.

Brazil, the biggest and most powerful South American country, played a very important role. ANATEL, the Brazilian Telecommunications Watchdog, went through a public consultation

---

[2] A guard band is a small band of about 2.5–5 MHz between dedicated frequency bands to protect the dedicated frequency bands against interference from each other.

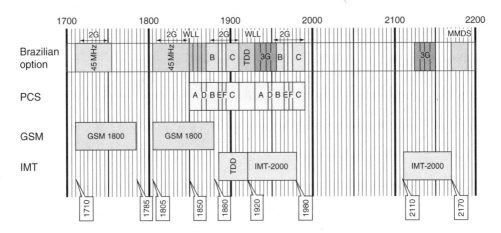

Figure 5.4 PCS, GSM and IMT-2000 spectrum plan for Brazil. *Source*: UMTS Forum's Latin America Group.

process on the 1800 MHz and 1900 MHz bands in late 1999 with a hearing on 16 December 1999. The UMTS Forum made a milestone-setting proposal (Eylert, n.d.), explaining how to integrate GSM and UMTS/3G into Brazil's future spectrum plan (Figure 5.4).

ANATEL's decision on this consultation was postponed until after the World Radio Conference in June 2000 to avoid influencing the WRC-2000 process. The subject was highly controversial, because a decision in favour of allocating GSM in this band would follow the European and Asia-Pacific format, but would be in contrast to the spectrum plans of other Latin American countries. Consequently, it would allow the operation of 3G/IMT-2000 in the WARC-92 allocated frequency bands, coordinated with Europe and Asia-Pacific, although this was not part of the consultation. Shortly after WRC-2000, ANATEL decided to follow the ITU recommendations and granted GSM operators spectrum in the 1800/1900 MHz band. That opened the door for the option to grant 3G spectrum in line with the WARC-92 agreement later, when the market matured.

5.1.3.4 Africa and Arab States

The African and Arab states followed the ITU discussions on IMT-2000 as agreed at WARC-92 in Malaga and WRC-2000 in Istanbul. Most of them have allocated the FPLMTS/IMT-2000 spectrum, but it seems to be that the markets in those countries are not ready for 3G yet. It is expected that from 2008 onwards these countries would offer 3G services, except a few countries around the Mediterranean and the Republic of South Africa, where 3G licences may be granted earlier.

5.2 WARC-92 SPECTRUM SHARING SCENARIOS

As soon as the WARC-92 had confirmed the allocation of spectrum for FPLMTS/IMT-2000 it became the responsibility of each country's regulators to determine the basis on which they would grant spectrum slices (frequencies) to operators. Because there were no equivalent examples to follow, the process was not easy. Which parameters should be considered in the process? Which factors would be the most important to take on board? Let us put ourselves for a moment in the shoes of a regulator and list the most relevant points together:

1. You must strive to create a fully competitive environment, which means deciding the minimum number of operators. Obviously, there must be at least two, but such duopolies have not been very successful in the past from the consumers' point of view. On the other hand, if the number is too large, there will be insufficient spectrum to allow the building of large networks, so no economies of scale, high costs and prices, and a risk of the market fragmenting into small and unattractive businesses from an investment point of view may occur.
2. When there are several competing technical standards, which of them (if any) should be recommended or made a condition of the licence? Mandating only one standard would secure national and international roaming, but puts the regulator in the difficult position of having to make a technology choice, something which governments are increasingly reluctant to do. On the other hand, is it better from a regulatory point of view to accept several technologies and leave it to the market to decide, which one(s) succeeds? This decision has e.g. an impact on guard bands and therefore on the efficiency of the spectrum usage.
3. What will be the landmass/population that should be covered and what kind of coverage will be required? For example, do you expect to cover only the main economic areas, which would be suitable for about a quarter of the population, or do you recommend a high population coverage of about 80% or even more up to 99.98%? Unless the system is a satellite network, in which coverage is instantaneous, you need to consider the evolution of the system over the life of the licence and how best to define when the different phases should be reached. Or can this be left entirely to the market? More of this aspect will be discussed in Chapter 6.
4. What are the market expectations? How many subscribers, subscriptions, services, applications etc. do you expect? Here, a market study is very helpful and the UMTS Forum has supported the industry and the administrations with a series of separate market studies as outlined in Chapter 2.
5. What kind of influence has the market evolution on the traffic characteristics for the network? In second-generation networks frequent measurements are taken of traffic characteristics during the busy hours and the quiet hours. Engineers try to optimise the network with this data, and marketing people investigate tariffs and other marketing tools to influence subscriber behaviour so that the traffic levels are smoothed out. Most of these GSM results were available to regulators, so that they could make decisions based on these numbers. As long as you expect symmetric traffic the model may be straightforward. When a specific uplink/downlink asymmetry has to be considered, it is much more complicated. As there are no advanced mobile networks in existence, this work is entirely new and so there is quite a bit of uncertainty in these assumptions.

This is not an exhaustive list, but merely an illustration of the problem faced by the regulators. As will be seen from the UMTS Forum studies discussed below, there are many other factors which affect the amount of spectrum needed.

In the following sections two traffic models of the UMTS Forum are described, one for the 3G spectrum allocated at WARC-92 and a second one for the so-called extension band

2500–2690 MHz, allocated at WRC-2000. The first traffic characteristics calculation concludes with up in some scenarios descriptions. The defined scenarios were tested against the first traffic forecasts based on market studies and previous UMTS reports. The Forum looked at eight deployment scenarios for UMTS and examined each of them for their viability in terms of the cell loading and levels of service capabilities for each of the hierarchical layers (see Section 5.2.9).

Based on the most recognised Forum's market studies (UMTS Forum, 2000a, 2001b,c, 2002a) a second traffic characteristics calculation has been undertaken (UMTS Forum, 2003b). This will also be discussed later in this chapter as well.

The UMTS Forum, as an international cross-industry body, investigated the minimum spectrum operators would need to run a viable mobile multimedia business in the initial phase of UMTS/3G. This was essential to give all the market players, administrations and operators some common basis on which they could formulate spectrum allocation policies.

When the early UMTS Forum's studies were produced in the late 1990s they were focused on the 15 Member States of the European Union (EU15) and many of these assumptions and therefore conclusions may well not be valid for countries outside of Europe, but could nevertheless serve as useful guidance.

This investigation into the minimum spectrum demand per UMTS operator relates to public terrestrial UMTS networks in the EU15 states and is based upon UMTS market and spectrum estimates made previously by the Market Aspects Group (MAG) and Spectrum Aspects Group (SAG) of the UMTS Forum (SAG, 1998). This first study with its recommendations has helped the regulators to make their regulatory decisions in an appropriate environment. The satellite spectrum discussion is not considered here because satellite turned out to be much less important than people thought in the early days of UMTS/IMT-2000. Readers who are interested in this area may look at publications of ITU on this subject or read report no. 6 of the UMTS Forum (1998d) in more detail.

In principal, UMTS and narrowband satellites to a certain extent may share the same market. It is important to observe that WRC-03 made a global allocation to the MSS (Earth-to-space) in the band 1668–1675 MHz and a global allocation to the MSS (space-to-Earth) in the band 1518–1525 MHz, i.e. an additional 2×7 MHz for MSS. However, satellite services have not developed as expected and most came under Chapter 11 creditor control/bankruptcy protection in the USA. Some have taken up niche markets after the Iraq War 2003 with some 100 000 subscribers in total, which is small compared to cellular services (*Wall Street Journal*, 2004).

5.2.1 FIRST TRAFFIC CHARACTERISTIC CALCULATIONS OF THE UMTS FORUM

In order to permit multimedia applications on a sufficiently large scale while conserving spectrum, both circuit- and packet-switched radio transmission technologies are assumed. The traffic calculations consider both transmission principles. Further, it is assumed that asymmetric traffic distributions may influence the spectrum demand on the uplink and downlink side and this is accommodated by designating different bandwidths.

The calculations of frequency spectrum requirements for UMTS cannot be limited to multimedia services and high-speed data services. A number of the foreseen UMTS services, such as voice and low speed data services, are today delivered by second-generation systems, and this will be the case also when UMTS has been introduced, at least for a transitional period.

As a first step, the calculations of frequency spectrum demand have been made taking into account all mobile services, excluding only services from fixed or quasi-fixed systems. The spectrum calculations concentrate on the European scenario. Scenarios in other continents have not been investigated in the same way, but some results from those areas of the world will be referred to later.

The estimation was built on the assumption that all mobile traffic, both from wideband and multimedia services as well as from narrowband services similar to the present second-generation services, is carried by UMTS. It was assumed that 90% of the total speech and low speed data traffic up to 2005 will be carried over existing second-generation networks, that 60% of the indoor traffic will be carried over licence-exempt networks, and that high (2 Mbps) and medium (384 kbps) multimedia services are packet services which are tolerant of delay. It is important to note that although the majority of users will continue to use speech most of the capacity is needed for multimedia services. As identified in the market study (UMTS Forum, 1999b), it was expected to continue to grow strongly after 2005 and additional spectrum would be required in the time after. These forecast numbers (users, traffic volumes etc.) constituted the basis for the spectrum estimates. The estimation included public as well as business and private users. It built on calculations made for the years 2005 and 2010.

For the year 2002, when UMTS introduction was initially planned, market conditions on their own did not determine the necessary amount of spectrum. To enable the provision of multimedia services in UMTS with a continuous coverage, a frequency band of about 2×20 MHz would be needed for public licensed use by each operator. This result can be derived from the RACE II projects ATDMA and CODIT (RACE, 1995). Systems presented as candidates for UMTS within ETSI (i.e. the RACE II FRAMES project) verified these estimates.

Recognising that in the initial phase limited spectrum may be provided for public systems, an additional band of 20 MHz would need to be designated as start-up band from the year 2002 for non-public non-licensed in-building low mobility systems. Such systems are seen as playing a key role in establishing a strong market for multimedia terminals and, more importantly, in stimulating a requirement for public access 'away from base'. In addition, more capacity will be freed for public systems. This UMTS Forum recommendation has proved to be true as can be seen by the setting up of WLAN services for in-building and neighbourhood use, which started broadly in Europe and the USA in 2003. However, this field is seen more as a niche market and will be discussed later in detail (see Chapter 6).

The spectrum requirements for such non-public non-licensed in-building low mobility systems are included in the total spectrum requirement estimates for 2005 and 2010. However, in 1999 the UMTS Forum thought that such systems would require separate frequency allocations of 20–40 MHz out of the estimated required spectrum in order to avoid interference with public systems. With the introduction of WLAN this approach has changed slightly.

It is expected that the forecast number of persons using mobile services is roughly equal to the number of people using voice service via mobiles. All other services are considered as supplementary to the basic mobile voice service. One person will thus be using several services. The forecasts mentioned earlier show that the use of the higher bandwidth multimedia services will increase over time, leading to a proportional decrease in the share of the voice services. However, it is not foreseen that the absolute volume of voice services will be diminishing. Due to the undefined nature of future services, particularly the multimedia services, a certain degree of care is needed in interpreting the estimated figures. The estimates of spectrum requirements are for the network busy hour, but the profile of traffic for each service type varies through the day. The future mix of services should result in spectrum being utilised more evenly than the present, particularly through the use of delay in high volume data applications.

Factors that are not treated in the calculations and which may further increase the required spectrum include higher traffic rates, higher penetration and user density variations. Nor are factors that might reduce the spectrum requirements, such as half-rate speech codecs, low-rate video codecs, adaptive and/or distributed antennas, efficient statistical multiplexing and overall improved carrier-to-interference (C/I) performance, considered in the performed calculations.

Improvements in technology will lead to improvements in spectrum efficiency. However, this potential may be partially reduced if improved quality is chosen, which is expected to be a market requirement. In addition, cellular radio has a practical difficulty in the problem of finding cell sites in optimal locations. This has first to do with the emission of microwave radiation, which people in some areas of the world fear. Second, CDMA technology does require different network planning compared to second-generation networks. This would lead beyond the scope of this book and the reader should refer to the specific literature (e.g. Kim, 1999; Walke, 2000).

5.2.2 SPECTRUM EFFICIENCY AND SPECTRUM SHARING

A significant element of the UMTS vision is the need to achieve a major improvement in spectrum efficiency compared to that already being achieved for second-generation mobile systems. The increases in spectrum efficiency will need to be found from four major sources:

- Radio transceiver technology, including access technology, modulation and coding, adaptive interference management, diversity techniques and smart antenna technology.
- Applications and services technology, including the use of packet transmission, asymmetry management, compression techniques and agent technology.
- Traffic management, especially via the use of delay management and tariffs to manage peak-to-mean traffic ratios.
- Radio channel access management, i.e. the management of instantaneous access to the spectrum, to reduce the probability of idle channels during peak traffic hours.

The two first points are entirely dependent on the choice of technology which is done by ETSI and other standardisation organisations, and these are therefore not in the regulatory field. The third point is in the commercial field and should not be regulated either.

In the context of the last point, it has been suggested that the sharing of a common pool of spectrum by operators might be a method of significantly improving spectrum efficiency, thereby minimising the overall demands for spectrum for UMTS. In this scenario, a pool of spectrum would be available that could be accessible to the various operators according to their current traffic demands. The idea of operator spectrum sharing was formally raised by ERO (1996). The UMTS Forum has examined the topic from technical, regulatory and commercial viewpoints.

More important in this context seems to be WLAN or Wi-Fi, as it is called in the USA (CNet News, 2004). The reader will be given a brief introduction to the spectrum implications for this technology as well, when the second traffic calculation is discussed. But it should be remembered that WLAN use spectrum in the ISM bands (2.4 GHz/5 GHz) and these bands are *not* part of the IMT-2000 spectrum! A more general discussion of WLAN/Wi-Fi is presented in Chapter 2 from the market perspective and in Chapter 4 from the technology perspective.

5.2.2.1 Technical Considerations

Sharing the same frequency spectrum between several operators could result in higher trunking efficiency and savings in guard bands. However, these savings do not take into account the airtime overheads which arise during call set-up, clear-down, handover etc. These overheads will not diminish due to spectrum sharing; on the contrary, the additional complexity of operating with shared spectrum may significantly increase the overheads. This could negate much of the perceived advantage of operator spectrum sharing.

The technical problems associated with operator spectrum sharing are likely to be significant. It would be especially difficult to control the quality and interference environments of a particular network, which are important factors in competitive differentiation. Dynamic channel allocation, with radios that seek interference-free channels in completely independent networks, is probably not the solution to this problem.

These technical problems lead to thoughts about synchronisation of networks or shared radio infrastructure. A logical further step is the single spectrum access operator, which would provide radio access across the entire spectrum using one set of base stations, and which would then sell airtime to the mobile network operators. From the spectrum efficiency point of view, this may be seen to offer some benefits, but the need to ensure fair and equal spectrum access by all mobile operators means regulatory and contractual restraints on individual operators. This might negate many of the sharing benefits. This idea has some similarities to a mobile virtual network operator (MVNO), which leases a complete network infrastructure from an existing licence holder and markets the complete service, including mobile terminals and billing arrangements, under its own brand. Such MVNOs have evolved on a limited scale, e.g. Virgin Mobile in the UK.

The sharing of spectrum between terrestrial and satellite UMTS networks will not generally be feasible, due to the expected wide differences in received power flux density and transmitted power levels between the terminals operating in these systems. Therefore, it is necessary to make separate spectrum allocations for terrestrial and satellite UMTS networks. However, feasibility of spectrum sharing between the UMTS satellite downlink component and indoor, unlicensed use needs continuing studies.

5.2.2.2 Commercial Considerations

An operator's maximum potential revenue is approximately proportional to the spectrum available. Therefore, access to a larger pool of spectrum would certainly be attractive to operators. To build a business case, UMTS operators must, however, be certain about the spectrum to which they have uninhibited access. One possibility is to give each operator a guaranteed minimum number of channels. This will reduce the spectrum efficiency advantage.

For the remaining free pool of spectrum, the operators will try various methods to attract users to their own networks. Operators may be reluctant to invest in radio hardware for this spectrum if there is no real certainty that the additional channels would be available when needed. Should they improve quality or services, or increase the number of base stations, then this is consistent with higher spectrum efficiency, but the cheapest way is to increase transmitter power levels. The commercial drivers may therefore lead to network developments that lower spectrum efficiency. This is in contrast with the situation in their own spectrum assignments, where the commercial drivers make them increase the capacity of the network and as a consequence also the spectrum efficiency.

One method of spectrum sharing is to allow users access to several or all of the operators in the same region. In this way, the users share the spectrum instead of the operators. While this method avoids some of the commercial problems of other sharing methods, technical and commercial problems remain. The user terminals have to ensure that the spectrum is efficiently utilised, which might increase the airtime overheads, and operators have to compete for users on a call-by-call basis. It is not yet clear if a stable market situation with investment incentives can be achieved in this situation.

5.2.2.3 Regulatory Considerations

Access by many operators to a common pool of spectrum will necessitate detailed regulation to prevent domination of the pooled resources by any single operator. Regulation will also be necessary to prevent operators from investments in technology that will diminish other operators' possibilities to use the spectrum. It is probable that the regulatory constraints will reduce the potential for spectrum efficiency gains.

5.2.3 SYMMETRIC AND ASYMMETRIC SPECTRUM USE

It is anticipated that UMTS high multimedia and medium multimedia traffic will be asymmetric in data flows between uplink and downlink, and the remainder of the traffic will be symmetric (see Table 5.10), leading to different requirements for radio spectrum in the two directions. However, before much can be decided about how the spectrum might be configured to take account of such factors, it is necessary to consider what is meant by asymmetry in this context.

The spectrum asymmetry can be defined as the ratio of transmitted downlink bits to transmitted uplink bits in a given integration time. The integration time is all-important as the picture changes depending on the observation window.

Within quasi-instantaneous time scales (say <10 s), all traffic, including speech, would undoubtedly be found to be highly asymmetric. In speech conversation generally only one party speaks at once. Even in picture messaging situations, visual activity is normally associated with the currently active participant. Messaging and file transfer services will clearly be asymmetric in these time scales. However, while the asymmetry will be of a high order within these short time scales, for some services (e.g. speech) the direction of the asymmetry will keep reversing.

Over the duration of a multimedia session (defined as the multimedia equivalent of a call), the session asymmetry can be very different to the quasi-instantaneous values. Integrated over these longer time scales, normal speech conversation would be considered symmetrical, as might videoconference sessions. However, messaging, file transfer and information-gathering transactions would continue to have a high degree of asymmetry in the information content.

Over a long period of time (day, week or month) and integrated over all customers using the UMTS network, there will be an overall net degree of asymmetry in the number of bits flowing in the uplink and downlink channels. This will be averaged over the mix of services being used by customers and the net balance of session asymmetry in the calls completed. It is this net degree of asymmetry that is primarily of concern to the spectrum planners. The net asymmetry is expected to be greater than unity, i.e. the totality of downlink traffic will require more spectrum capacity than the totality of uplink traffic.

The development of the UMTS spectrum requirements within the UMTS Forum is based on six market environments, ranging from the central business district (CBD) to rural. For any of the defined UMTS service groups (simple messaging, high multimedia etc.) the degree of asymmetry in the traffic being generated in each of these environments may vary, because the tasks that people will want to do with UMTS will vary.

The comments above relate to the degree of asymmetry within the end-user traffic. Transactional asymmetry can be very high at this level. The UMTS Forum figures for medium multimedia and high multimedia services are ~40:1 and 200:1, respectively. However, the transmission of this information over a mobile network requires that additional system information be added to cope with packet transmission, error handling and protocol overheads. These additional overhead signals will have a proportionally greater effect on the low data rate direction of an asymmetrical traffic flow, and will have the effect of reducing the overall asymmetry. Simple estimates based on fixed network hardware capabilities suggest the actual worst case asymmetry might be about 10:1 (e.g. ADSL technology). Studies based on the World Wide Web indicate similar ratios (supported by an investigation in Nieminen, 1996) but other studies dealing with other market segments may indicate other ratios, which may be closer to unity. This can be called the network-level asymmetry and the spectrum will need to be tailored to this requirement. A more detailed discussion of traffic asymmetry can be found in Hewitt (1998), Huber (1998) and France Telecom Mobiles (1998). Traffic asymmetry has been allowed for in these calculations and therefore should not be a problem in the initial years of the UMTS rollout.

5.2.4 SERVICE CLASSES AND SWITCHING MODES

In its report 'Spectrum for IMT-2000', SAG (1997) converted the UMTS multimedia applications described in the Analysys/Intercai (1997) report, which were based on a variety of future applications for data users, into a limited number of UMTS service classes (Table 5.1). These service classes allow spectrum calculations independent of the various applications the user may have. They define the basic capacity requirements for the UMTS spectrum calculations. Response and delay time requirements from such applications were not taken into account. However, they will affect the mode of operation on the air interface and will consequently impact its efficiency.

The asymmetry factors in Table 5.1 are taken from SAG (1998). The service classes have different characteristics. High interactive multimedia services, e.g. video telephony, require isochronous transmission, and also switched data and speech services and they are therefore

Table 5.1 UMTS service classes and characteristics

Services	User net bit rate (kbps)	Coding factor	Asymmetry factors	Effective call duration (s)	Service bandwidth[a] (kbps) UL/DL	UMTS switch mode[b]
High interactive MM[c]	128	2	1/1	144	256/256	CS
High multimedia (MM)[d]	2000	2	0.005/1	53	20/4000	PS
Medium multimedia[d]	384	2	0.026/1	14	20/768	PS
Switched data	14.4	3	1/1	156	432/43.2	CS
Simple messaging	14.4	2	1/1	30	28.8/28.8	PS
Speech	16	1.75	1/1	60	28.8/28.8	CS

[a] The service bandwidth is the product of columns 2, 3 and 4.
[b] CS, is circuit switched; PS, packet switched.
[c] Characterised by high speed data rates, symmetric and reasonably continuous transmission and minimum of delays.
[d] Characterised by moderate data rates, medium to large file sizes, asymmetric and bursty transmission and tolerance to a range of delays. GPRS and HSCSD come into this category.
© UMTS Forum (1998c).

calculated as circuit-switched services. This means that the average call duration time corresponds to the actual connection set-up time. The effective call duration depends on the occupancy factor, which is e.g. 0.5 for speech and 0.8 for video telephony. For packet-switched services, the call duration is calculated as the sum of time intervals, where data are actually transferred via the air interface. Thus, the occupancy factor in this scenario is equal to 1 (Figure 5.5).

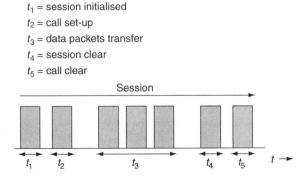

Figure 5.5 Packet transmission over UMTS air interface. *Source*: UMTS Forum (1998c).

The earliest market forecast of the UMTS Forum (UMTS Forum, 1997, 1999b; Analysys/ Intercai, 1997) shows 16% multimedia users for the year 2005 and 30% for the year 2010. These figures are low in contrast to speech. However, the picture becomes different when the traffic volumes are considered. Here, multimedia dominates by far over speech in the year 2010. This can be seen in the traffic calculations of the SAG report on UMTS spectrum estimates (1998). It can also be seen there that multimedia traffic has high asymmetry regarding uplink and downlink. In this context asymmetry has to be seen as an overall net degree of asymmetry in the number of bits flowing in the uplink and downlink channels (Hewitt, 1998).

The UMTS spectrum available in the year 2002 must be sufficient to handle the UMTS traffic up to the year 2005 in order to give operators confidence in the possibilities to develop and introduce UMTS.

5.2.5 UMTS MARKET ASSUMPTIONS

The UMTS Forum's spectrum estimates from the year 1997 show, that speech and low data rate services can remain in second-generation networks, if all second-generation spectrum identified for GSM and DECT is allocated. In this context it should be noted that new entrants, which do not have an existing second-generation network, may also gain a UMTS licence, in which case a higher focus on providing speech and low rate data services as well as multimedia may occur, e.g. on GPRS using existing networks as MVNO. The existing second-generation operators who gain a UMTS licence may also have a high interest in providing speech services because of a lack of capacity in their second-generation networks. It has been assumed that the traffic is shared equally between all the operators. It has also been assumed that 90% of the total speech traffic will remain on second-generation systems in the initial phase. This is difficult to estimate, especially in the light of dual-mode GSM/ UMTS terminal development, which will allow for a long time the combined use of GSM and UMTS networks. Therefore, in the first part of this study, it has been assumed that most speech services as well as low speed data services will be switched via GSM radio networks in the respective GSM spectrum.

This assumption of 90% of speech and low data on existing GSM networks recognises the existing success of GSM and that this will continue. It must also be remembered that GSM will probably continue to offer the better wide area coverage for speech and low data users up to at least the year 2005.

It should be noted, however, that the 10% assumption is for up to the year 2005. After 2005 it can be expected that the number of subscribers using only speech and low speed data will reduce in favour of subscribers taking up the whole range of applications from speech to multimedia.

5.2.6 TRAFFIC CAPACITY REQUIREMENTS IN URBAN ENVIRONMENTS

In the spectrum calculations for UMTS six potential user environments were considered (UMTS Forum, 1998d; SAG, 1998). The analysis of the population distribution in Europe shows that 50–60% of the population are in the urban area (Figure 5.6). Further, the peak of spectrum demand comes from the urban pedestrian environment.

Only the urban environments (CBD, pedestrian and vehicular) are considered now for calculating the spectrum requirement for the UMTS radio network as it is expected that the

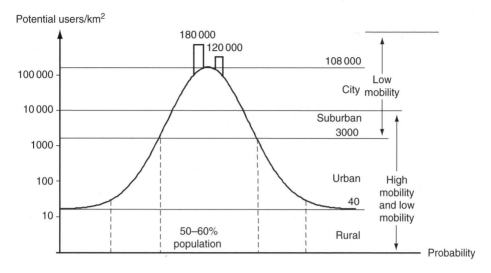

Figure 5.6   Distribution of population in Europe. © UMTS Forum (1998c).

highest traffic densities, and consequently the highest bandwidth requirements, are in dense urban areas. It is assumed that 60% of the in-building traffic originates from licence-exempt networks (Huber, 1998). Hence only 40% of the traffic forecast for CBD is considered in this calculation. The calculation uses the assumptions from SAG (1998). The result for traffic per km^2 (aggregate traffic before adjustment for QoS parameters) is shown in Table 5.2.

Table 5.2   Traffic calculation for UMTS services

Service class[b]	Aggregate traffic in the busy hour (Mbps/km^2)[a]			
	Year 2005			
	Uplink		Downlink	
	CBD (40%)	Urban	CBD (40%)	Urban
HMM 2 Mbps	0.15	0.1	30.6	22
MMM 384 kbps	0.06	0.05	2.5	1.8
HIMM 128 kbps	1.1	0.4	1.1	0.4
Speech/low speed data	2.5	2.3	2.5	2.3
Sum	3.8	2.85	36.7	26.5
All environments	6.65		63.2	

[a] Aggregate traffic includes the net bit rate, coding factor, uplink/downlink factor, and a 20% signalling overhead. For CBD the cell size is smaller than that for the other environments. Therefore an equivalent traffic value based on the cell sizes of the urban pedestrian and vehicular environment is used in order to simplify the conversion into spectrum demand in Table 5.3.
[b] HMM, High multimedia; MMM, medium multimedia; HIMM, high interactive multimedia.
©UMTS Forum (1998c).

5.2.7 MINIMUM SPECTRUM PER OPERATOR FROM A TECHNICAL POINT OF VIEW

This section states the simplifications and approximations used in the spectrum calculations and defines some deployment scenarios. The general assumptions are as follows:

- UMTS market forecast and traffic volumes according to market study for EU15 states (Analysis/Intercai, 1997): split into packet-switched and circuit-switched traffic;
- the estimation from a market point of view is based on the figures available from UMTS Forum (1997, 1998d), Analysys/Intercai (1997) and SAG (1998);
- for the years 2002–2005, services carried over UMTS networks will be mainly focused on multimedia. Speech + low speed data are used mainly in second-generation systems, with 10% of the speech and low data rate services carried in the UMTS bands (i.e. 90% on second-generation systems);
- three radio environments determine the spectrum requirement: urban pedestrian, urban vehicular and central business district/indoors. Suburban and rural are not critical in this respect;
- ETSI Air Interface standard according to high level UTRA requirements (ETSI, 1998): 5 MHz carrier spacing; frequency division duplex (FDD) and time division duplex (TDD) modes;
- the maximum available data rate is 384 kbps for macro- and micro-cells with full mobility and 2 Mbps for micro- and pico-cells with low mobility;
- each operator has exclusive spectrum (i.e. no frequency sharing);
- market shared equally between all operators. This assumption ignores the fact that an operator could have a greater share of the market. It was felt impractical to use other assumptions, but this fact should be taken into account;
- fixed hierarchical cell structures are used with traffic assigned to specific layers;
- only spectrum demand for public operator licensed use is considered;
- 155 MHz is available for terrestrial UMTS as identified in CEPT ERC/DEC/ (97)07 on the introduction of UMTS, i.e. 1900–1980 MHz, 2010–2025 MHz and 2110–2170 MHz (ERC, 1998).

Several assumptions have been made regarding the capabilities of the technology, the size of the market and the development and take-up of new services. Some of these assumptions may turn out to be pessimistic or optimistic. Tele-traffic models for multimedia networks carrying mixed data rate traffic of both circuit- and packet-switched services were not included. The calculations initially derive the traffic loading per carrier without any allowance for QoS.

A dedicated explanation of how to evaluate whether a certain scenario is adequate to satisfy the overall traffic demand can be found in annex 2 of UMTS Forum (1998c). It also contains a calculation of the cell capacity in relation to the QoS. Based on that calculation a QoS factor of about 3 seemed to be sufficient to allow for acceptable blocking of circuit services and reasonable delay constraints on packet-switched services.

The following assumptions are made:

- The technology will support a channel spacing of 5 MHz. This channel spacing includes all necessary guard bands assuming the same spectrum efficiency. It is therefore assumed that 12 FDD and 7 TDD carriers are available.

Table 5.3   Maximum available data rates

Cell type	Mobility class	Maximum available user net bit rate
Macro	High	384 kbps
Micro	High/low	384 kbps/2 Mbps[a]
Pico	Low	2 Mbps

[a] 2 Mbps (low mobility only) may be possible close to the base station.
© UMTS Forum (1998c).

- The maximum available data rate for a carrier in FDD mode (2×5 MHz) is 2 Mbps (uplink/ downlink) capacity for low mobility or 384 kbps (uplink/downlink) for full mobility. The maximum available data rate for a carrier in the TDD mode (5 MHz) is 2 Mbit/s uplink or downlink (but not simultaneously) for low mobility applications.
- Cell types, mobility classes and maximum data rates (user bit rates) are assumed according to Table 5.3.
- Scenarios assume that 60% of CBD traffic is carried in licence-exempt networks.
- CEPT in Europe has made available all ITU spectrum except 15 MHz which are already used for DECT. This results in 155 MHz of spectrum for terrestrial services with an additional 60 MHz identified for UMTS satellite services within the 2 GHz MSS bands.
- High and medium multimedia are assumed to be tolerant of reasonable delay .

It should be noted that:

- If the spectrum efficiency factor is higher than assumed the spectrum per operator could be less.
- The minimum economic cell size may be larger than assumed, particularly for micro-cells. In this case the spectrum requirement per operator could be increased.
- The packaging options evaluated here are based on a fixed hierarchical structure. Other possible deployment scenarios could lead to further packaging options, e.g. it may be possible in the future to reuse the same carrier for different cell layers, such as the same carrier to support an FDD macro-cell outdoors and a FDD pico-cell indoors.
- If the spectrum demand proves to be greater than that derived within UMTS Forum (1998c), this could lead to either a need for more total spectrum for a given number of operators or a reduction in the number of operators per country for a given spectrum allocation.

5.2.8 SPECTRUM ALLOCATION AND DEPLOYMENT SCENARIOS

In order to realise the benefits of the information society as soon as possible, and for UMTS to be a mass market global solution, every effort will be needed to make UMTS as commercially attractive as possible. To do this requires that operators can make the optimum use of the scarce resource of spectrum. They should be given the maximum flexibility to cater to the possible different customer/market requirements. UMTS will make use of advanced cellular structures aimed at maximising network capacity. Such network structures are generally called hierarchical cell structures and can be implemented according to the specific operator's needs.

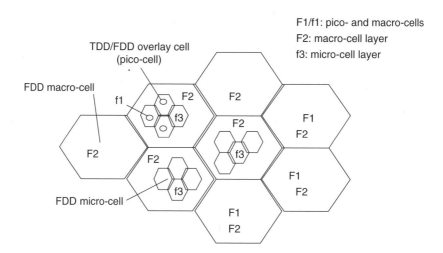

Figure 5.7   Flexible UMTS three-layer network. © UMTS Forum (1998c).

For an optimal UMTS radio network, it is proposed that UMTS is to be planned using a hierarchical cell structure (Figure 5.7), using micro-, macro- and pico-cells. Consideration needs to be given as to which element of UTRA (FDD or TDD) should be employed for which type of cell. Below is shown a possible use of this hierarchical cell structure, where high traffic demand could require 3 layers. With flexible deployment it could be possible in some locations for an operator to re-deploy picocell channels for macro cells outside of urban areas.

All CBD HMM traffic is assigned to the pico-cells, whereas all urban-pedestrian HMM traffic is allocated to micro-cells (partly at a lower data rate) and all HMM urban-vehicular traffic is carried by macro-cells at a lower data rate.

The FDD macro-cell provides the wide area coverage and is also used for high-speed mobiles. The micro-cells are used at street level for outdoor coverage to provide extra capacity where macro-cells could not cope. A cluster of them is shown here, although they could be deployed singly. It would seem likely that these micro-cells would not be hexagonal in shape but canyon-like, reflecting the topography of a street and be perhaps 200–400 m in distance. This would be very specific to the city type.

The pico-cell would be deployed mainly indoors, in areas where there is a demand for high data rate services such as laptops networking or multimedia conferencing. The way in which these pico-cells could be deployed would be dependent on their maximum range in given environments (indoor and outdoor). Such cells may be of the order of 75 m in distance. A limiting factor will be the range of these terminals when used for high data rate services. Given the high demand, this will impact battery capacity.

As a basis for a network operator an allocation of paired frequencies for FDD operation is necessary:

- 2×5 MHz will allow a single layer only; a hierarchical cell structure is not feasible in this case;
- 2×10 MHz gives room for a two-layer structure, e.g. a macro-cell layer together with either a micro-cell layer or pico-cell;

- 2×15 MHz allows the deployment of a complete hierarchical cell structure where the traffic demand is high, or a mix of layers such as one macro-cell and two micro-cells;
- 2×20 MHz allows increased flexibility and additional capacity.

In addition to the allocation of paired frequencies, an operator may need an allocation of unpaired frequencies for TDD operation, in particular for low mobility applications indoors:

- 5 MHz may be required in order to give satisfactory capacity for asymmetric traffic;
- 10 MHz would give more flexibility and additional capacity for asymmetric traffic.

5.2.9 POSSIBLE SCENARIOS

Eight scenarios were selected to study their ability to cater for the projected demand for UMTS services. These eight varied from 2×5 MHz to 2×20 MHz paired + 5 MHz unpaired spectrum. Table 5.4 defines the eight scenarios with the different allocations per operator. The total predicted traffic is assumed to be equally shared between each operator.

In addition, it should be noted that other countries and regions in the world will probably have different market demands and requirements, therefore other scenarios for minimum spectrum per operator may also be relevant.

Eight deployment and allocation scenarios have been introduced, which have been discussed from a technical point of view. Based on these scenarios, the maximum number of operators for each scenario is derived in Table 5.4, taking into account the spectrum availability situation in Europe. In addition to the frequencies allocated in the different scenarios, there may be some spectrum of that decided in the ERC Decision in reserve, depending on national constraints. From the analysis of the market situation (UMTS Forum, 1999b), it was concluded that a technical deployment with the assumptions made in Alcatel *et al.* (1998) and SAG (1998) is not possible. For this reason all scenarios are examined taking into account the results of spectrum allocation and deployment scenarios from the previous section, assuming different parameters for cell sizes and spectrum efficiency figures than used in SAG (1998).

Table 5.4 Deployment scenarios

Scenario	Paired frequencies allocated to one operator (MHz)	Unpaired frequencies allocated to one operator (MHz)	Maximum number of operators	Traffic per operator Mbps/km² uplink	Traffic per operator Mbps/km² downlink	Traffic per operator Mbps/km² total	Spectrum not allocated (MHz)
1	2×5	—	12	0.55	5.3	5.85	35
2	2×5	5	7	0.7	9.2	10.1	50
3	2×10	—	6	1.1	10.5	11.6	35
4	2×10	5	6	1.1	10.5	11.6	5
5	2×15	—	4	1.6	16	17.6	35
6	2×15	5	4	1.6	16	17.6	15
7	2×20	—	3	2.2	21	23.2	35
8	2×20	5	3	2.2	21	23.2	20

© UMTS Forum (1998c).

Using these values for each scenario, the different services and environments are assigned to cell layers. It is then evaluated whether it is possible to carry the traffic with the deployed scenario. The methodology is described in more detail in annex 2 of UMTS Forum (1998c). The loading of the different cell layers should be below 100% if the scenario is feasible to carry the traffic as required, with some provision for quality of service. The cell sizes (Table 5.5) are selected as close as possible to those values from SAG (2004). Improved spectrum efficiency figures are obtained based on the results of the ETSI simulations (ETSI, 1997a,b). As the most critical situation is in dense urban areas, only the three environments CBD, urban-pedestrian and urban-vehicular are considered. All assumptions and figures used for this calculation are included in annex 2 of UMTS Forum (1998c).

5.2.9.1 Scenario 1

The Federal Communications Commission (FCC) in the USA requires the minimum bandwidth per public operator to be 2×5 MHz. With this limitation UMTS will allow the deployment of a one-layer network. Based on these parameters, the UMTS Forum's calculation, outlined in Table 5.6, shows the consequences of such an allocation.

Furthermore, in order to be able to deliver the 2 Mbps service in combination with all other types of services in an actual traffic situation, two layers of micro-cells are necessary. One

Table 5.5 Assumed base station distances and cell areas

Cell type	Distance (km)	Cell area (km^2)
Macro	1	0.288
Micro	0.4	0.138
Pico	0.075	0.005

© UMTS Forum (1998c).

Table 5.6 Effective cell loading with flexible distribution of traffic between network layers with QoS factor of 3

Scenario	Macro layer		Micro layer		Pico layer	
	Load DL (%)	Load UL (%)	Load DL (%)	Load UL (%)	Load DL (%)	Load UL (%)
1	NA	NA	249	31	NA	NA
2	NA	NA	54	33	14	6
3	NA	NA	63	38	15	1
4	56	45	52	30	16	7
5	60	45	82	49	23	1
6	48	35	42	25	25	11
7	55	41	57	34	31	1
8	55	41	38	23	33	15

NA = not available.
© UMTS Forum (1998c).

layer has to be reserved for the 2 Mbps service; the other serves other customers. It will probably be necessary to temporarily remove all other users from the carrier in the vicinity of the 2 Mbps user. This makes a second micro-cell layer necessary. In this scenario with only one carrier no 2 Mbps service can be delivered. The flexibility to serve different kinds of users at the same time will also be diminished. It will also be very complicated and expensive to achieve seamless coverage without a macro-cell layer.

Scenario 1 is not feasible from a technical point of view, as it will not provide the full range of UMTS services. Without a macro-cell layer it will be virtually impossible to cater for fast moving mobiles, and make it very difficult for UMTS operators to provide coverage comparable with that expected by GSM customers today in rural and urban areas. With the assumed cell size and operating on only one layer the capacity will not be sufficient to carry the heavy traffic load as forecast for the EU15. Therefore this scenario is not recommended.

5.2.9.2 Scenarios 2 and 3

Even if the addition of a pico-cell layer increases the capacity to more or less the required level, the problem of delivering the full range of UMTS services in all areas remains. The pico-cells will have only a limited coverage, indoors and in hotspots. While these scenarios can cater for the traffic load predicted for the EU15 countries, they still suffer from not having a macro-cell layer.

Without a macro-cell layer it will be virtually impossible to cater for fast moving mobiles, and make it very difficult for UMTS operators to provide coverage comparable with that expected by GSM customers today. Therefore these scenarios are not recommended.

5.2.9.3 Scenario 4

This scenario can handle all traffic. Scenario 4 would work with five operators as well as with six operators, because there is the required 20% spare capacity. That means that each operator has sufficient spare capacity to handle 20% extra traffic with the spectrum assumed for each in this scenario, which would enable five operators to carry all the traffic, even though six operators are assumed here.

However, having only one channel per cell layer may not provide the flexibility that is required. There may be problems delivering high data rate services in some areas.

Due to the disadvantages of this scenario, it is not a preferred solution.

5.2.9.4 Scenario 5

The deployment situation is similar to the situation in scenario 4, except that paired bands instead of unpaired bands are used for pico-cells. However, the situation is worse because the total spectrum assumed to be allocated to operators is less than in scenario 4. The loading on the micro-cell layer (downlink) is increased significantly over all scenarios (except 1). This scenario also does not have the benefit of TDD spectrum, which makes the efficient handling of asymmetric traffic difficult. Again, as in scenario 4, this solution may lack in flexibility. Due to the disadvantages of this scenario, it is not a preferred solution.

5.2.9.5 Scenario 6

This scenario is the preferred solution for the minimum spectrum required by a public UMTS operator.

The scenario with full functionality that occupies the least amount of spectrum per operator is scenario 6. It allows one macro-cell layer, two micro-cell layers and one pico-cell layer.

A possible distribution of carriers and of traffic between the carriers is shown in Table 5.7. Pico-cells have been built out to cover enough area to take up 25% of the traffic, which probably means not only hotspots and some indoor areas, but also dense town centres.

If a more limited deployment of pico-cells is desirable from an economic point of view, more spectrum for an extra micro-cell carrier has to be found.

5.2.9.6 Scenarios 7 and 8

These scenarios also allow full functionality and have enough capacity to carry all traffic.

5.2.9.7 Summary and Recommendations

Scenarios 1–3 are not practicable as they do not provide the full functionality required.

Scenarios 4–8 can be considered further. The choice of which scenario should be preferred cannot be the same for all countries and markets. There will be a tradeoff between the cost of rolling out a network, since extra spectrum (all other things remaining equal) should lead to a less costly rollout, and the benefits of having extra operators. This is not, however, a technical issue, and is therefore not discussed here. There are also issues concerning the flexibility with which an operator can deliver services and the realisation of a global mass market.

However, the preferred solution is 2×15 MHz paired and 5 MHz unpaired for the minimum spectrum required by a public UMTS operator.

It is envisaged that by the year 2005 techniques will be available which will allow UMTS to exploit the spare capacity in paired channels due to traffic asymmetry. This will give an

Table 5.7 Possible traffic distribution between cell layers

Service class	Share (%)		
	Macro-cell layer	Micro-cell layer	Pico-cell layer
High MM	0	0	100
MMM	5	70	25
HIMM	5	70	25
Switched data	10	65	25
Simple messaging	10	65	25
Speech	10	65	25
	1 FDD carrier	2 FDD carriers	1 TDD carrier
Loading of downlink	48	42	25
Loading of uplink	35	25	11

© UMTS Forum (1998c).

Table 5.8 UMTS service capabilities

Scenario	Paired frequencies allocated to one operator (MHz)	Unpaired frequencies allocated to one operator (MHz)	UMTS service capability
1	2×5	—	Limited
2	2×5	5	Limited
3	2×10	—	Limited
4	2×10	5	Some possible restrictions
5	2×15	—	Some possible restrictions
6	2×15	5	Full
7	2×20	—	Full
8	2×20	5	Full

© UMTS Forum (1998c).

operator the required margin to continue catering for demand for some time past the year 2005. Sooner or later there will, however, be a need for more spectrum for UMTS.

The aim of carrying out this analysis is to give guidance for the minimum spectrum demand per public UMTS operator in the initial phase. Eight scenarios are calculated and evaluated based on the market forecast for EU15. The results of the evaluation suggest the UMTS service capabilities shown in Table 5.8.

Based on the assumptions in this study, including the practicality of distributing traffic loading relatively evenly between the hierarchical cell layers, and assuming that the use of TDD in the unpaired spectrum is the more efficient way of handling asymmetric traffic, the UMTS Forum considers that scenario 6 (2×15 MHz + 5 MHz) is the preferred minimum.

However, depending on country-specific situations regarding, for example spectrum, operators, market, asymmetry and traffic, other spectrum allocations per operator may be more appropriate. Therefore, depending on spectrum availability, more paired or unpaired spectrum could be allocated to an operator if higher traffic demands it.

The 2×15 MHz (scenario 5) would be technically sufficient to allow a UMTS service to start up and offer the full range of services envisaged at this time, but may not provide a flexible deployment of hierarchical cells.

It has been shown that from a purely technical point of view the minimum spectrum requirement (by definition) is 2×10 MHz (FDD) + 5 MHz (TDD) (scenario 4). This scenario provides sufficient capacity to carry the projected traffic for Europe and the full range of UMTS services, but may not provide a flexible deployment of hierarchical cells. There may be problems delivering high data rate services in some areas.

Scenarios 7 and 8 allow full functionality and have enough capacity to carry all traffic.

5.3 MORE SPECTRUM NEEDED: EXTENSION BANDS FOR UMTS/IMT-2000 AND WRC-2000 RESULTS

This section reflects the results of the UMTS Forum's work on extension bands that has been done within its Spectrum Aspect Group (SAG), providing an industry view on the suitability of these candidate bands for the UMTS/IMT-2000 terrestrial component. It takes into account some of the work which has been undertaken in the relevant ERC and ITU-R groups.

In their efforts to identify additional spectrum for the UMTS/IMT-2000 terrestrial applications, administrations called for justification and evidence to support the preparatory work for the ITU WRC-2000. As an European industry response, the UMTS Forum summarised the relevant information on spectrum issues, listed the candidate extension bands and gave information on the usage and benefits for third-generation applications. It also explored examples of allocation scenarios in order to support the WRC-2000 decisions on agenda item 1.6.

The timing of the requirement for additional spectrum within individual countries differed, may continue to differ and will depend on the development of the market in those countries. It has been calculated that administrations will need the full additional spectrum between 2005 and 2010, which has been accepted by the WRC 2000 Conference (Hewitt, 2000).

It has been widely recognised that UMTS/IMT-2000 should not be just an improved version of today's second-generation mobile services. It should be really a new concept, offering new and significant opportunities driven by data applications and market demand. This new concept has led to a higher spectrum demand as higher bit rate services need much wider bandwidths than 'traditional' voice applications. However, US operators are already using the spectrum and communicate on mobile broadband. The latter could easily be understood that they see mobile broadband communications just as an improvement of today's GSM/GPRS/EDGE network, an argument which you cannot ignore completely.

A large part of the activity performed on UMTS/IMT-2000 in different international forums has been concentrated on the spectrum domain. Consolidation of spectrum requirement results is ongoing worldwide, taking into account geographic disparities and influences, market and traffic impacts as well as technical and system aspects.

A longer term strategic approach to UMTS/IMT-2000 spectrum aspects is necessary to prevent a possible spectrum shortfall during successful development of UMTS/IMT-2000. The UMTS Forum considered the securing of the additional spectrum for UMTS/IMT-2000 terrestrial mobile applications to be a major objective for WRC-2000. Successful UMTS/IMT-2000 deployment will depend not only on the market needs and the technological progress, but also on the timely availability of spectrum. The full potential of the third-generation networks can be realised only if the frequency allocations are appropriate.

The UMTS Forum provided for the WRC-2000 conference an industry view on the suitability of a list of candidate bands, taking into account the regulatory aspects. The Forum considerations included the background on existing usage, the advantages that each band offers to UMTS/IMT-2000, the potential availability of the band, and the likely risks that this availability might not be realised in time combined with some implementation aspects.

UMTS Forum SAG considered the whole issue with a view to determining the UMTS Forum position, taking into account:

- the results of studies on characteristics, properties and current usage of candidate bands;
- the technical and operational issues related to sharing and coordination with other services that need to be coordinated within CEPT;
- the conclusion on the possible worldwide and regional spectrum most appropriate to the needs of third-generation systems from an industry point of view;
- frequency planning aspects of the possible allocation scenarios.

5.3.1 WORLDWIDE MOBILE MULTIMEDIA MARKET

In its effort to locate additional spectrum for the UMTS/IMT-2000 extension bands adminis-trations called for justification and evidence to support the preparatory work for the ITU WRC-2000 agenda item 1.6, and the UMTS Forum contributed a report on minimum spectrum for operators (1998c) extensively describing the calculations and projections.

The world market for mobile multimedia services was projected on the basis of the current status of market development compared to the EU. It may be interesting to note that markets outside the EU may dominate the world market. The Asia-Pacific region showed an upturn in the past few years and it will most likely be in a lead position by the year 2010, according to the UMTS Forum's market studies (2000a, 2001b,c, 2002a).

The information in figure 5.8 shows a high level of growth extending until at least the year 2010. Although less marked in the case of North America, the similarity of the trends suggests that the requirements for additional spectrum will be broadly similar in most principal market areas.

Regarding mobile traffic and capacity, the expected enhancement of telecommunications and the drive from multimedia and packet-switched applications were considered in particular. Based on this, among others, the UMTS Forum concluded that about 582 MHz will have to be identified for the highest traffic areas in the year 2010. The requirement includes the bands currently designated for second-generation systems and the UMTS/IMT-2000 core bands, plus new spectrum. This leaves a requirement of 187 MHz for additional spectrum in Europe.

5.3.2 TOTAL SPECTRUM ESTIMATES

The first estimates of the total spectrum requirement of 582 MHz for mobile terrestrial systems in year 2010 were done by SAG in December 1997. The additional amount of spectrum needed

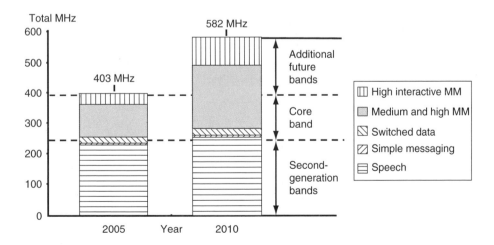

Figure 5.8 Terrestrial spectrum estimates for the years 2005 and 2010 (including second-generation). © UMTS Forum (1999a).

for UMTS/IMT-2000 to meet market needs was estimated to be 187 MHz in year 2010. These estimates were confirmed in UMTS Forum (1998d).

The methodology developed by the UMTS Forum has been adopted – with slight modifications – by CEPT administrations within ERC/TG1 and presented to ITU-R/TG 8-1. On this basis ITU-R/TG 8-1 has elaborated a generic methodology which could also be used for other public mobile radio systems. This methodology is applicable to both circuit-switched and packet-switched traffic and can accommodate both symmetrical and asymmetrical services.

An ITU-R/TG8-1 common understanding of input assumptions as well as agreed projected traffic forecasts have given the basis for the derivation of the initial results of a global spectrum estimate for the UMTS/IMT-2000 terrestrial component. These results, which are in line with those given by the UMTS Forum, are applicable to all regions.

The UMTS Forum spectrum requirement calculated for terrestrial third-generation services is based on:

- market forecast and penetration;
- potential user density in different environments;
- service and traffic characteristics;
- infrastructure and technical factors.

As of February 1999 the current total spectrum figures – quoted by the main international bodies considering this issue – for the year 2010 are as given in Table 5.9.

On the assumption that the core band will provide up to 170 MHz for third-generation mobile, depending on existing regional allocation, a maximum additional 150–309 MHz would be required from 2005 to satisfy the projected market until year 2010. As much as possible of this additional spectrum must be harmonised on a global basis. The results presented in Table 5.9 confirm the trends of spectrum requirement worldwide.

Table 5.9 Summary of spectrum requirements calculation

	Total spectrum needs (MHz)	Spectrum already identified for second and third generation (MHz)	Additional spectrum (MHz)
UMTS Forum (1998c)	582 [a]	395	187
CEPT administrations (ITUR, 1998a)	556	395	161
US administration (ITUR, 1998c)	499	190	309
Japanese administration (ITUR, 1998d)	440		
ITU-R (Draft) (ITUR, 1998e)	532		

[a] 403 MHz in 2005.
© UMTS Forum (1999a).

5.3.3 BASIC PRINCIPLES INFLUENCING THE CHOICE OF FREQUENCY BANDS

The vision for UMTS/IMT-2000 is a mobile communication environment that can deliver voice and high data rate digital and multimedia services to a worldwide mass market. As such, it will require spectrum allocations that are, as far as possible, harmonised on a global basis. That could be achieved via the Decisions taken at WRC-2000, as you will see later.

Important considerations are the minimising of costs and the stimulation of competition in the supply of mobile services to this mass market. The industry (operators and manufacturers) believe these will best be achieved if UMTS/IMT-2000 spectrum is concentrated in as few separate frequency bands as possible, and with sufficient spectrum in each band to allow several competing operators to have allocations with broadly similar propagation characteristics and hence radio network costs.

Preference should be given to options that permit contiguous spectrum. If this is not possible, then additional bands as close as possible to the existing core band would have advantages.

While this is a perfectly valid assumption, it is worth considering the benefits of other frequency options. Propagation conditions are not the same at all frequencies. At the lower (UHF) frequencies, a greater range is possible for a given transmitter power, giving wider area coverage, and signals more easily penetrate through vegetation, into buildings. The lower frequencies are also more able to maintain service over rough terrain. Therefore there may also be benefits in having access to some lower frequency allocations to allow UMTS/IMT-2000 to be made available in rural areas, difficult terrain and in the lower population areas where longer ranges and better signal penetration would be a clear advantage, e.g. the 700 MHz band in the USA.

UMTS/IMT-2000 networks starting in time period 2002–2005 will be rolled out in the 2 GHz 'core band', because that will be the available spectrum. The additional spectrum requirement is expected to arise as the peak demand in urban areas increases. It is unlikely that an operator could or would restructure its network (cell sizes, frequency reuse, antennas) in rural areas if it subsequently gained access to frequencies which have advantages in coverage in such areas. To take advantage of the benefits of lower frequencies for implementing UMTS/IMT-2000 in rural and/or developing areas, such frequencies would therefore need to be available at a fairly early stage. Nevertheless, not all countries will implement UMTS/IMT-2000 simultaneously, and the availability of lower frequency extension bands could provide the key to a cost-effective and technically attractive deployment in developing countries and less populated areas where a 2 GHz roll-out is commercially unattractive.

5.3.4 A FLEXIBLE APPROACH TO ADDITIONAL IMT-2000 SPECTRUM

In recognition of the general principles discussed above, it is quite useful that the objectives at WRC-2000 identified a minimum number of relatively large IMT-2000 extension bands and to ensure these have primary Mobile Service allocations, preferably in all three ITU Regions. These bands were identified for IMT-2000 applications, on a similar but not necessarily identical basis to footnote S5.388:

S5.388: The bands 1885–2025 MHz and 2110–2200 MHz are intended for use, on a world-wide basis, by administrations wishing to implement the Future Public Land Mobile

Telecommunication System (FPLMTS). Such use does not preclude the use of these bands by other services to which the bands are allocated. The bands should be made available to FPLMTS in accordance with resolution 212 (Rev WRC-95).

National administrations can now organise on the national basis total IMT-2000 allocations in order to match their specific market needs from within the 'core band' and the identified extension bands. Administrations will not necessarily make all the bands or all of any such bands available, so long as the required total was met from within the identified bands, and was made up from individual allocations of at least the minimum size suggested above.

This approach offered administrations the advantages of flexibility in implementing IMT-2000 alongside existing services. It provided clear guidance for industry by indicating, at an early stage, the frequency ranges within which equipment would need to operate. Finally, it gave operators and investors confidence in the medium- and long-term growth potential of UMTS/IMT-2000.

As noted above, additional spectrum was required for both the FDD and TDD modes of UMTS/IMT-2000, and in the former case due attention has been paid to the minimum duplex spacing requirements of low-cost terminals and base stations. By default, the core band offers a duplex spacing of 190 MHz, but it is recognised that flexibility in duplex arrangements might be attractive, and will be essential when considering the different extension bands. It is not therefore essential that a single duplex spacing is preserved for all future UMTS/IMT-2000 allocations.

5.3.5 THE UMTS FORUM ANALYSIS OF EACH CANDIDATE EXTENSION BAND FOR UMTS/IMT-2000

The purpose of UMTS Forum report no. 7 (1999a) was to bring together all of the relevant information on the global allocations for the possible candidate extension bands for UMTS/IMT-2000. Studies in that report were based on previous results of the UMTS Forum's reflection on extension bands (1998d) as well as ITU/R TG8–1 analysis (ITUR, 1998b).

The following bands were discussed within the draft CPM text:

- 470–806 MHz
- **806–960** MHz
- 1429–1501 MHz
- **1710–1880** MHz
- 2290–2300 MHz
- 2300–2400 MHz
- **2520–2670 MHz**
- 2700–2900 MHz

(The bands indicated in bold were supported by CEPT.) A detailed discussion on all bands is given in UMTS Forum (1999a). In the following discussion, we concentrate on the final chosen extension bands 806–960 MHz, 1710–1880 MHz, 2300–2400 MHz and 2520–2670 MHz, all taken from report no. 7 (UMTS Forum, 1999a).

5.3.5.1 Analysis of the Candidate Band 806–960 MHz

Background

Replacement of analogue broadcasting by digital television (with better spectrum efficiency, e.g. 3 or 4 programmes per 8 MHz channel, improved frequency reuse) in the band below 862 MHz may offer the possibility of considering a part of this spectrum for UMTS/IMT-2000. Parts of the band are used also for other services, and demand for these services is likely to continue.

A large part of this band, i.e. above 862 MHz, is currently used extensively for existing mobile systems e.g. GSM900 (880–915/925–960 MHz) and could be refarmed in the longer term.

Benefits of this Band for UMTS/IMT-2000

- Already a global mobile allocation.
- Potential for longer range coverage due to the more beneficial propagation conditions would in particular benefit rural areas, areas of low population density and developing countries. However, a longer range can best be realised on a FDD basis.

Availability and Risks

The upper part of the broadcasting is used for the introduction of Digital Video Broadcasting – Terrestrial (DVB-T) in many countries. If the band is used for IMT-2000, it would conflict with DVB-T. Coordination of the analogue broadcasting plans is subject in Europe to the Stockholm agreements (1961). These have been complemented by the Chester agreements (1997) for digital television. Coordination is thus likely to be necessary between a country using analogue television in the band and another country using IMT-2000. It is necessary to change the Stockholm/Chester agreements in order to introduce other systems in this band.

At the end of the coexistence period between analogue and digital TV services, the removing of analogue TV would make some spectrum available only if a rearrangement of TV assignments should occur. This might be a long process.

If a part of this band is selected for IMT-2000, then an evolution of the existing applications is necessary.

- If those existing applications are licensed mobile services, then there is a possibility to transfer to IMT-2000. Availability of this band for UMTS/IMT-2000 can only be made progressively in the longer term as existing use decreases.
- If those applications are non-mobile, sharing may be possible on a geographical basis.

Time scales for availability of this band for IMT2000 may differ.

Almost the whole band 806–960 MHz is used by second-generation mobile systems in different regions of the world. There is therefore an opportunity to find bands that would be appropriate for IMT-2000. Since the availability is not the same in every country, the whole band 806–960 MHz would be identified worldwide for IMT-2000 on a regional basis at the beginning but with the important potential to become a global allocation later.

The existing second-generation allocations in different regions give an indication of the possible availability of 50–89 MHz of spectrum as follows:

- 50 MHz in Australia;
- 70 MHz in Europe;

- 80 MHz in the USA, 806–821/851–866 MHz and 824–849/869–894 MHz. Current use: US cellular. The lower part coexists with the European proposal, too;
- 86 MHz in Canada;
- 89 MHz in Japan.

Summary
- The GSM part of the whole band could be refarmed for third-generation applications in the longer term (according to the market evolution).
- This band was identified on the ITU level to migrate later from second-generation to IMT-2000.

5.3.5.2 Analysis of the Candidate Band 1710–1880 MHz

Background
A large part of this band, i.e. 1710–1785/1805–1880 MHz, is currently used for existing mobile systems, e.g. GSM1800 in Europe, and could be refarmed in the longer term.

Benefits of this Band for UMTS/IMT-2000
- It is already a global mobile allocation.
- Close proximity to the core band.
- Similar propagation characteristics to the core band.

Availability and Risks
Different band plans for cellular mobiles (PCS and GSM1800) are currently employed in different countries. Countries are likely to want to retain the same band plan for IMT-2000.

Summary
- This band was identified on the ITU level to migrate later from second-generation to IMT-2000.

5.3.5.3 Analysis of the Candidate Band 2300–2400 MHz

Background
Fixed service and aeronautical telemetry.

Benefits of this Band for UMTS/IMT-2000
- Wide enough band for the additional band of UMTS/IMT-2000.
- Mobile allocation in all three Regions.

Availability and Risks
- Some countries have indicated this spectrum could be made available.
- In some countries the aeronautical telemetry services are implemented but there are only a few transmitters. The aeronautical telemetry services need bands below 3 GHz, their requirement is from 40 to 110 MHz and the harmonisation is suitable.
- Other systems like wireless cameras and ENG/OB operate in this band.

- Some difficulties are possible because of the other users (e.g. S-DAB is developed in USA and could be spread, new military fixed links in some countries with new equipment, radiolocation).
- In ERC/TG1 this band is no more considered as an European candidate.

Summary
- This band was identified only for China to solve their extending spectrum problems for mobile broadband services in the longer run, but not specified as an IMT-2000 extension band.

5.3.5.4 Analysis of the Candidate Band 2520–2670 MHz

Background
Fixed service, electronic news gathering/outside broadcasting (ENG/OB) and multipoint distribution applications.

Benefits of this Band for UMTS/IMT-2000
- Allocation to mobile services in all Regions.
- The band is wide enough to cover most of the forecast UMTS/IMT-2000 spectrum requirement.
- Might be paired with 2290–2300 MHz and 2300–2400 MHz (2.3.5 and 2.3.6) for even greater opportunities.

Availability and Risks
- In the USA and Canada multipoint distribution applications are used. However, in Europe such systems are of very limited use.
- Part of the band, 2520–2535 MHz/2655–2670 MHz, may be used for MSS in some countries. The same part can still be used for terrestrial services in other countries.
- Geographical sharing (urban/rural) might facilitate the transition or even enable, in the longer term, remaining operation of other services.

Summary
This was the most probable candidate for additional band globally. It offers 150 MHz, possible for most of the CEPT administrations and in all ITU regions, starting from year 2008 or later subject to market demand, after phasing out of existing usage.

It also offers at least the following possibilities:

- 150 MHz TDD band;
- 2×70 MHz paired (maximum, if duplex-separation allows);
- 1×120 MHz is possible in 2535–2655 MHz (if the MSS allocation is left out) for TDD or FDD, 2×110 MHz could be possible if paired with 110 MHz within 2290–2400 MHz or 2700–2900 MHz.

5.3.6 DECISIONS MADE AT WRC-2000

The main results of the conference are taken from Hewitt (2000) and are summarised here.

The existing 'core bands' were protected. The identification of the existing IMT-2000 'core' bands was successfully defended, and the status of these bands remains unchanged.

The band 2500–2690 MHz was identified for IMT-2000 in all three ITU Regions (Region 1 – Europe (including Russian Federation), Africa and the Middle East; Region 2 – The Americas; Region 3 – The rest of the World). This will be the main expansion band for Europe for the terrestrial component of IMT-2000. Initially only 2520–2670 MHz will be available as the other 2×20 MHz sub-bands are allocated to the mobile satellite service (MSS). Some countries in Asia-Pacific will also use all or part of this band for IMT-2000 expansion. Some Region 2 countries (CITEL) may study this band for possible longer-term use as expansion spectrum for IMT-2000, perhaps on a shared basis with other applications, but it is not available in most CITEL at the present time because of extensive MDS usage. Details of the identification of this band can be found in footnote S5AAA and Resolution COM5/24 (WRC-2000). The modifications for all the new IMT2-2000 bands are given in the frequency table in article S5 of the Radio Regulations.

The band 1710–1885 MHz was also identified for IMT-2000 in all three Regions. This is the preferred IMT-2000 band for the CITEL administrations (except the USA). It will in fact be their 'core band' as the original IMT-2000 bands have been used for PCS etc. However, as the band is also identified on a global basis, it will allow eventual evolution of GSM 1800 to IMT-2000 at a later stage, if required. Some Asia-Pacific countries with broadcast interests in 2500–2690 MHz will also use this band for IMT-2000 extension spectrum. Details of the identification of this band can also be found in footnote S5.AAA and Resolution COM5/24 (WRC-2000).

First and second generation mobile bands below 1 GHz (specifically in 806–960 MHz, e.g. GSM 900 and US cellular) are also identified, on a worldwide basis, for IMT-2000 to provide administrations with an opportunity to migrate these bands to IMT-2000 in the longer term. Details of the identification of these frequencies can be found in footnote S5.XXX and in the Resolution COM5/25 (WRC-2000).

The USA achieved recognition that it may wish to implement IMT-2000 in its recently auctioned bands at about 700 MHz, and China secured recognition that 2300–2400 MHz will be its preferred (national) expansion band. However, these bands are not specifically identified for IMT-2000.

The satellite component of IMT-2000 also received appropriate attention. The MSS bands at 1980–2010 MHz and 2170–2200 MHz are already identified for IMT-2000 applications by virtue of their location within the previously identified core bands. WRC-2000 decided also to identify the other MSS bands below 3 GHz for IMT-2000 to allow an equitable opportunity for all the MSS satellite operators to provide IMT-2000 services via their satellite networks. Details can be found in footnote S5.SSS and Resolution COM5/26 (WRC-2000).

Finally, in an innovative decision, the conference agreed further provisions (a final footnote and resolution) that now enable high altitude platform stations (HAPS) to be used in the existing core bands as platforms for base stations for the terrestrial component of IMT-2000 (see footnote S5.BBB and Resolution COM5/13 (WRC-2000)).

5.3.7 CONSEQUENCES FOR THE FUTURE OF IMT-2000

The new provisions in the Radio Regulations provide full flexibility for administrations to choose whether or not to implement IMT-2000 in these bands. Those administrations wishing to go ahead with IMT-2000 can decide on how much spectrum to make available, which of the identified band(s) they want to take the spectrum from, and when they want to make it available. For administrations that have less immediate interest in IMT-2000, the use of the identified bands for other applications within the services to which the bands are formally allocated is not precluded in any way. This flexibility reflects the uncertainties over the size and timing of individual markets for IMT-2000 (especially in the developing countries).

With a clearly defined spectrum environment, the manufacturers within the three Regions now know the limits of the frequencies for which the terminals must be designed. By having a limited number of globally identified bands, the manufacturers have the best opportunity to reduce costs via economies of scale.

One of the most important attributes of IMT-2000 will be the capability for global roaming with a single terminal, allowing people to do 'anything, any time, anywhere' in their day-to-day telecommunications context. The WRC Decisions to identify a limited number of new bands on a worldwide basis greatly facilitate global roaming with low-cost terminals.

This was an excellent result for the UMTS Forum, and it now provides ample opportunities for enterprising operators to build and grow global IMT-2000 businesses.

The Decisions of WRC-2000 are very significant for the future of IMT-2000, which is one of the ITU's biggest ever projects. Taken together, the decisions create an excellent radio spectrum environment within which IMT-2000 networks can be deployed and developed to the benefit of all peoples of the world.

5.4 THE SITUATION AFTER WRC-2000

Table 5.10 gives an almost complete overview on spectrum availability for mobile communications in the 800/900, 1700–2200 MHz and 2.6 GHz bands after WRC-2000. It illustrates the different technical systems in these bands, the number of subscribers and operators by end 2002 and, as far as known, the future plans for these systems.

5.4.1 HOW TO USE THE NEW SPECTRUM 2.520–2.670 MHZ

Immediately after WRC-2000 a discussion took place on how to use the new spectrum. Although this spectrum should be coordinated worldwide, all regional groups have different views. Operators, administrations and manufacturers all have their own ideas on how to split the band for FDD or TDD use. It should also be noted that some administrations are considering reallocation of current MSS spectrum to terrestrial use (e.g. the USA) because of a lack of MSS use. Additionally, the US FCC has made a decision that for the 2.6 GHz band, there will not be MSS and the whole 2500–2690 MHz band is for terrestrial.

Again, the UMTS Forum took the lead in devising possible solutions by building on the experience gained from the first traffic calculations. The UMTS Forum started a new study under the leadership of its SAG, called '3G offered traffic characteristics' (UMTS Forum, 2003b),

Table 5.10 Spectrum availability and global mobile subscriber figures for terrestrial and satellite services, end 2002.

IMT-2000 band (MHz)		Band	System	Subscribers millions (end of 2002)	Services	Operators (end of 2002)	Future plans[a]
806–960	MS	810–826	PDC	AMPS + other analogue: 30.0	Voice, data up to 384 kbps	467	
		824–849/ 869–894	AMPS / TDMA / CDMA	TDMA (aggregate): 109.2			
		880–915/ 925–960	GSM 900	CDMA (aggregate): 142.7			
		901–941/ 940–956	GSM / PDC[b]				
1710–1885	MS	1710–1785/ 1805–1880	GSM 1800	GSM (aggregate): 787.5			
1885–1980	MS	1850–1990	GSM 1900 / TDMA / CDMA	PDC (aggregate): 60.1			
				TOTAL: 1129.5			
1900–1920/ 2010–2025	MS (UMTS TDD)					103	
1920–1980/ 2110–2170	MS (UMTS FDD)		3G/IMT-2000[c]	W-CDMA: 0.153[d]		119 (3G Americas)	
1525–1544/ 1626.5–1645.5	MSS (GSO)					10	Data rates up to 432 kbps
1545–1559/ 1646.5–1660.5	MSS (GSO)			0.468[e]	Voice, data up to 144 kbps		

Frequency band (MHz)	Service	Subscribers	Data		Future plans
1610–1626.5/ 2483.5–2500	MSS (N-GSO)	0.147	Voice, data up to 9.6 kbps	2	New ICO
1980–2010/ 2170–2200	MSS (satellite UMTS, N-GSO)				2 × 3.5 MHz, plans in USA/FCC, S-DMB concept (first generation)
2500–2520/ 2670–2690	MS and MSS (UMTS)	0.028[f]		2	Terrestrial UMTS, S-DMB concept (second generation)
2520–2670	MS (UMTS)				Terrestrial UMTS

[a] In the frequency bands where there are as yet no subscribers, the future plans are indicated.
[b] PDC also uses 1429–1441 MHz/1477–1489 MHz.
[c] CDMA-2000 subscriber figures included in CDMA figures above.
[d] 1 million in August 2003.
[e] Does not include Volna and Solidaridad subscriber figures.
[f] The Japanese NSTAR a, b, c satellites had a total of 28 000 subscribers in March 2002. No information on INSAT subscribers was available.
Source: UMTS Forum.

published in November 2003 and publicly available on the Forum's website. The following section discusses the study and summarises the results.

5.4.2 3G OFFERED TRAFFIC ASYMMETRY

The scope of the above mentioned UMTS Forum report (2003b) was the determination of the probable aggregate offered traffic asymmetry in the overall busy hour based on the best available data on the characteristics of possible 3G services and the expectations, i.e. of how different users could use these services. It did not attempt to translate the aggregate traffic asymmetry into spectrum asymmetry as this step involves many additional considerations.

The report used the same modelling and methodology *ATIVA Research Tools®* as in the market studies (UMTS Forum, 2000a, 2001b) summarised in section 2.3.4, undertaken by Telecompetition Inc, and considered an environmental scenario representing a typical European high-density user environment with regard to population, geography, demographic characteristics, propensity to buy services, market scenarios etc. This environment is not therefore country specific.

The work was focused on the traffic characteristics within this environment in the year 2012 with the expectation that the nature of the UMTS traffic will have stabilised by that time. It was assumed that any growth in UMTS traffic beyond that time would be within the same general asymmetry framework.

The report is based on the projected characteristics of the 'UMTS Forum Services', i.e. the following six generic service categories that were identified and developed in UMTS Forum (2000a, 2001b):

- mobile intranet/extranet access
- customised infotainment
- multimedia messaging service (MMS)
- mobile Internet access
- location-based services
- simple voice and rich voice.

An assumption is that these remain a reasonable description of the types of services that will be provided over 3G mobile networks. It should be noted that other categorisation schemes are possible but would not change the conclusions. The key factor here is that all currently envisaged 3G traffic has been taken into account and, on the other hand, has not been counted more than once.

The report addresses on a country level the average service mix based on these six service categories with about one third (36%) business and two thirds (64%) consumer subscribers. Active subscribers within a smaller geography such as a single cell will vary significantly. The report is focused on the 'busy hour' traffic for the services, as this determines the overall spectrum requirements. However, for the multi-service environment it was recognised that the busy hours for each service might not – indeed, probably would not – coincide. Also, it should be noted that the report addresses traffic on 3G networks expected in the year 2012.

Figures 5.9 and 5.10 and Table 5.11 show the daily total traffic, the offered traffic asymmetry and the diurnal traffic load distribution for each service category. The total traffic load in these cases is based on assumptions regarding subscriber numbers, usage rates as well as

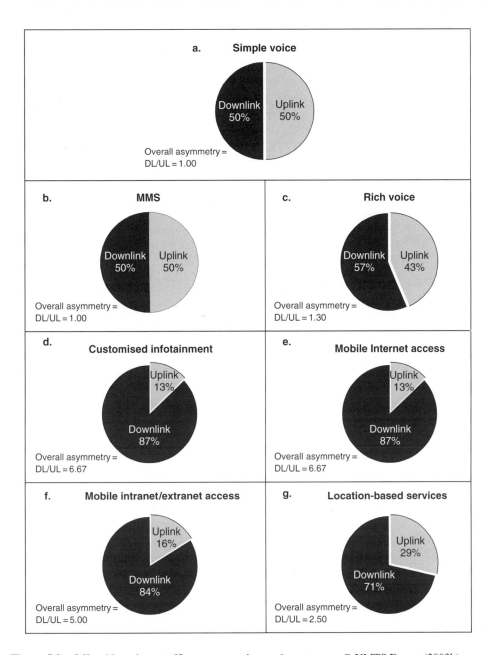

Figure 5.9 Offered busy hour traffic asymmetry by service category. © UMTS Forum (2003b).

activity and subscription types made for a typical European high-density user environment. Network overhead traffic for operation, administration and maintenance was not considered.

Figures 5.11 and 5.12 show the offered traffic asymmetry and the diurnal traffic, load distribution aggregated over all six service categories with and without simple voice traffic, respectively. To derive the aggregate offered traffic asymmetry in the overall busy hour the

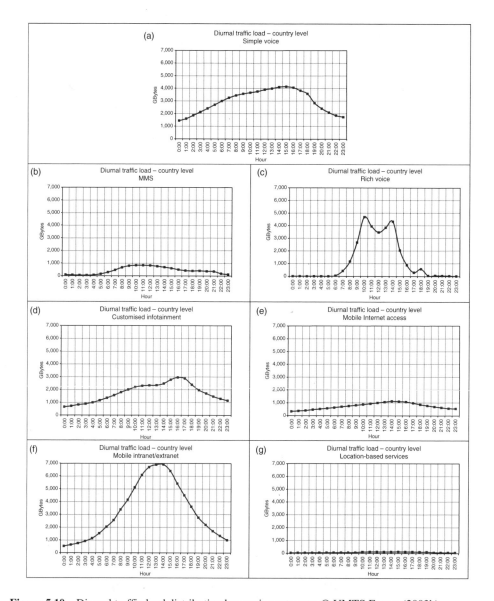

Figure 5.10   Diurnal traffic load distribution by service category. © UMTS Forum (2003b).

traffic asymmetry of each service category was weighted with the corresponding total traffic. Not shown here, but included in the report, was an analysis of coincident service busy hours. The asymmetry resulting from this worst case analysis supported the overall conclusions shown in Figures 5.11 and 5.12. The report concludes that the aggregate offered 3G traffic on a country level will tend to be asymmetric towards the downlink with a DL/UL asymmetry of about 2.3.

Table 5.11 Daily total and busy hour traffic by service category

Service category	Total daily traffic for the representative country (Tbytes)	Total traffic in service category busy hour (Tbytes)
Multimedia messaging service	9.92	1.02
Rich voice	28.66	4.73
Customised infotainment	42.04	2.95
Mobile Internet access	17.87	1.13
Mobile intranet/extranet access	78.14	6.70
Location-based services	1.30	0.09
Simple voice	71.28	4.26
Total	249.21	20.88

© UMTS Forum (2003b).

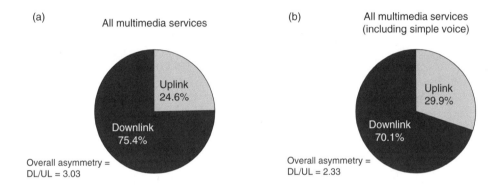

Figure 5.11 Aggregate offered traffic asymmetry in the overall busy hour. © UMTS Forum (2003b).

5.4.3 MEASURED ACTUAL GPRS AND WLAN TRAFFIC ASYMMETRY

Information on measured traffic data has been received from several operators, members of the UMTS Forum, mainly from their GSM/GPRS services. The sources are not identified in the UMTS Forum's report (2003b) for confidentiality reasons.

5.4.3.1 GPRS

In the first case GPRS traffic characteristics were measured over a period of 5 weeks in July–August 2003 (3 weeks of the period coincide with summer holidays) in two major city areas in one European country. Measurement data for workdays and weekends were averaged for every hour of the day and normalised to the maximum value observed. All data represents the bits per second user traffic without signalling. The results are shown in SAG (2004).

Similarly, measurements of GPRS traffic characteristics were conducted by a second operator in another European country in a major city area in June 2003 (4 weeks). These

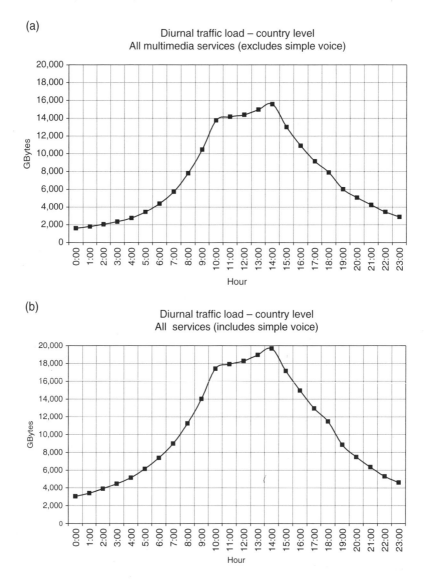

Figure 5.12   Aggregate diurnal traffic load distribution. © UMTS Forum (2003b).

measurements include service user data (without differentiation between services) and sign-aling traffic. The results are shown in SAG (2004).

A third operator measured its GPRS traffic in the entirety of a European country during 4 weeks in August–September 2003, following the main summer holiday season. In the measurements of uplink and downlink traffic, separate data was obtained for data-only and control traffic. The results are shown in SAG (2004).

It is considered that the control/signalling traffic in relation to the data-only at high traffic may be neglected in studies of asymmetry. From the measured actual GPRS traffic data of the three operators, the following may be noted:

- Workdays: there is a peak asymmetry a few hours around midnight, with the DL/UL ratio in the range of 2.5–3.5. However, at peak traffic the DL/UL ratio is typically 2–2.5.
- Weekends: the peak asymmetry is less during weekends, but the DL/UL average ratio typically remains 2–2.5 during the hours with relatively high traffic.

5.4.3.2 WLAN

Measurements during periods from early 2002 until spring 2003 by two different public WLAN operators in Scandinavia reveal that there is a slight DL/UL asymmetry in public WLAN services. This traffic was generated by users with laptop PCs at hotspot access points. The median DL/UL ratio during this period is close to 2.5 in both networks The results are shown in SAG (2004).

5.4.3.3 Summary of Measured Actual GPRS and WLAN Traffic Asymmetry

From the measurements of traffic data received so far, it may be concluded that traffic asymmetry may be expected within a range of DL/UL ratio between 2 and 3.5. At peak traffic hours the ratio tends to be somewhat lower, mostly below 2.5 (Figure 5.13).

5.4.4 SUMMARY AND RECOMMENDATION OF SECOND TRAFFIC CALCULATION OF THE UMTS FORUM

The study concludes that the aggregate offered 3G traffic on a country level will be asymmetric towards the downlink with a DL/UL asymmetry of 2.3. However, higher or lower asymmetry

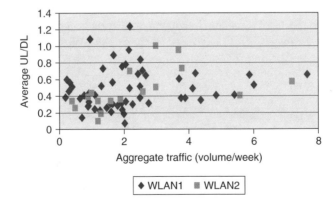

Figure 5.13 Average UL/DL ratio in WLAN as a function of aggregate traffic. © UMTS Forum (2003b).

could be expected at a cell level. Additionally, actual traffic asymmetry in the order of 2–3 was observed on GPRS and WLAN networks. While the actual and offered traffic asymmetry can not be directly compared at this point, the indications are that future asymmetry at a country level will be less than previously anticipated.

As traffic asymmetry is the most important parameter when determining the required spectrum asymmetry, this study is a good basis for the development of a suitable frequency arrangement for the 2500–2690 MHz band.

According to its tradition, the UMTS Forum recommends that:

- the information contained in the Forum's report (2003b) should be used as the basis for the decisions to be taken by ECC PT1 regarding the development of a suitable frequency arrangement for the 2500–2690 MHz band;
- these studies do show a need for more DL traffic capacity than UL traffic capacity on a country level;
- for the total 3G traffic on a country level a DL/UL asymmetry of 2.3 should be adopted as the working assumption.

Furthermore, as outlined in another UMTS Forum document (SAG, n.d.), regarding the frequency arrangement in the 2500–2690 MHz band, it recommends:

- FDD UL spectrum should be defined in the band, i.e. in-band pairing is supported. This excludes scenarios 4 to 7 of ITU-R M.1036–2. The preferred scenarios 1–3 are described in Figure 5.14.
- The size and place of the FDD UL block should be defined. Note that further discussion is required on which other points need to be fixed.
- A gap between the FDD UL block and the FDD DL block should be defined. A minimum value of 30 MHz was defined by 3GPP in TR 25.889 and the same value of 30 MHz was adopted by the Forum, noting that a larger gap could be beneficial for FDD operation and from the equipment point of view.

This is not yet the end of the story. Much work is still to be done in groups such as the UMTS Forum's SAG or the ECC PT1, until an international solution is agreed at ITU level.

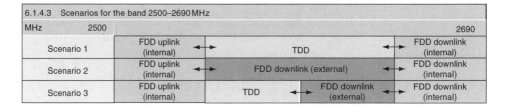

Figure 5.14 Scenarios for the band 2500–2690 MHz. *Source*: ITU-R M. 1036–2.

5.4.5 *FUTURE ACTIVITIES*

The UMTS Forum plans to investigate the translation of aggregate traffic asymmetry into spectrum asymmetry in a separate study. However, this study may not be available for the 2.6 GHz band plan discussions.

Further Forum studies will also address the modelling of 2G and 2.5G offered traffic and its migration to 3G networks. The overall service mix at a country level masks significant variations at the level of individual cells. Such spatial variations and their effects on traffic asymmetry are the subject of further investigation. This study will also determine any synergistic relationship between the Telecompetition and the European Project Momentum models. One topic is to study the implications of packet data transmission and mechanisms to evolve current circuit-switched traffic models to a packet-based analysis.

5.5 SPECTRUM BEYOND 3G AND PREPARATION FOR WRC-07

Although 3G is in its introductory stages today and the new allocated spectrum band of WRC-2000 is not defined yet in detail, the industry is already considering the next steps in mobile

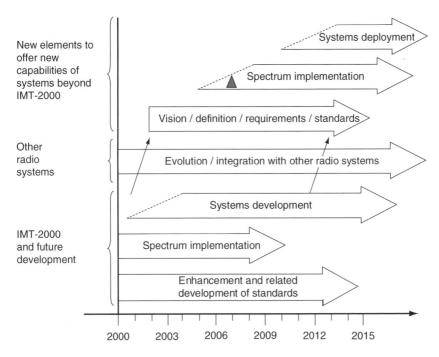

The dotted lines indicate that the exact starting point of the particular subject can not yet be fixed.

▲ Spectrum identification assuming that WRC-07 approves WRC-10 agenda

Figure 5.15   Timetable for IMT-2000 and beyond. *Source*: ITU-R WP8F.

communications. When GSM started in the early 1990s administrations and industry were already well on their way to defining and elaborating FPLMTS, later IMT-2000. Today, a similar situation is happening and although the industry is presently in a difficult economic situation it is desirable to start research work beyond 3G. Note that this need is for *research* and not product development or network deployment! At the time of writing, WRC-07 is only two years away and according to the general procedures in ITU the community has to agree on the agenda for WRC-10 at that conference! The subject will be discussed in more detail in Chapter 8. However, at this point some remarks are appropriate on possible action to be taken to move the discussion on 3G and beyond ahead. Figure 5.15 gives the timetable as suggested by ITU-R WP8F. Some more aspects can be found in Chapter 8. However, we will have to leave this topic at this point, as discussions are ongoing in most international forums and may not be clarified until shortly before WRC-07.

6

Regulation, Licensing and Global Terminal Usage

6.1 REGULATION

Why do we need regulation? Twenty years ago most countries of the world relied on their Post, Telegraph and Telephone Authority (PTT) to run all the communication services on behalf of the state as a department of the government. Investment in the copper infrastructure was funded by a combination of taxation and tariffs set by the PTTs, which operated as a monopoly. The same PTTs also owned and managed the radio spectrum on behalf of the state, and it seemed logical that this monopoly behaviour should be extended to the first public mobile communications networks.

6.1.1 HISTORICAL BACKGROUND

For many people it is difficult to understand why states should have the right to grant licences and distribute frequencies. Throughout history, societies have developed codes of conducts which members of society have accepted and agreed to follow for a successful working community. As a part of this evolution, states on behalf of all communities took over the right to distribute scarce resources to the benefit of society. Radio spectrum are also a scarce resource and so they were handled in the same way. On an international level, it was agreed to grant the ITU similar control. Under its rules member states distribute radio frequencies through national licences.

A rising influence on telecommunications policy and market was taken by regional and multilateral organisations. First of all, consider the inter-governmental trading body, the World Trade Organization (WTO). It has historically sought to lower barriers for tradable goods. In the 1980s, WTO extended its mandate to services negotiations including telecommunications.

The Mobile Multimedia Business: Requirements and Solutions   Bernd Eylert
© 2005 John Wiley & Sons, Ltd

In 1994, trade ministers established a Negotiating Group on Basic Telecommunications (NGBT) and in February 1997 an agreement was reached, and entered into force a year later. As a result, 69 countries made commitments to liberalise their telecommunication market under protocol 4 of the General Agreement on Trade in Services (GATS) of the WTO. Not all agreed to an immediate opening of their markets, but at least they agreed to inform about the status in their telecommunications markets and set a timetable for liberalisation (World Trade Organization, 2002). Nationally, the US government was the first which put in place a Telecommunications Act in 1984, followed by the UK Government in the same year.

6.1.1.1 Regulation in the USA

The US Congress passed legislation to break up the Bell/AT&T telephone system monopoly by the Telecommunication Act 1984 followed by the Telecommunication Act 1994. The Act's primary function is 'to provide a pro-competitive, deregulatory framework designed to rapidly accelerate private sector deployment of advanced telecommunications and information technologies and services to all Americans, by opening up all telecommunications markets to competition' (Mouritsen, 2002).

6.1.1.2 Regulation in the UK

In the UK, the Telecommunications Act 1984 set the framework for a competitive market for telecommunications services by abolishing BT's exclusive right to provide services. In the early 1990s the market was further opened up and a number of new national Public Telecommunications Operators (PTO) were given licences. This ended the duopoly that had existed in the 1980s when only BT and Mercury were licensed to provide fixed line telecom networks in the UK.

6.1.1.3 Regulation in the EU

The European Union (EU) exerts considerable influence on the telecommunication policies of its member countries. In fact the EU's influence goes beyond its formal members and has affected not only the new members which joined the EU on 1 May 2004, but also those which may come in later or are associated to the EU. They all aligned their telecommunications polices with the EU guidelines. The EU was instrumental in encouraging the introduction of competition when digital mobile cellular networks were launched in its Member States in the early 1990s. Its most significant achievement was the imposition of competition across all telecommunication market segments of its members as of January 1998. Four new EU Directives covering Framework, Authorisation, Access and Interconnection, and universal services were agreed in March 2002 with the aim of further developing a pro-competitive regulatory framework (European Commission, 2002). They were implemented on 25 July 2003. The emphasis of these Directives is on light touch regulation, technology neutrality and greater consistency across Europe.

6.1.2 REGULATION CONSEQUENCES AND POLITICAL IMPLICATIONS

The de-nationalisation (or privatisation) of fixed telecommunications in the USA showed the way to promote competition and, with it, the need for regulation and licensing becomes a more and more important role for administrations and the old and new operators. With the liberalisation of the telecommunications market the old PTTs were split into a regulator part and an operator part. Former old friends became opponents because one part became responsible for regulating the market whereas the other part became a pure operator whose main objective was to hold onto its dominant position in the market and to remain profitable. That was not easy in the beginning and caused some frustration on both sides. However, times have changed and all players have settled into their new roles.

Liberalisation and privatisation stimulate competition but in a former monopoly market, regulation is essential to give newcomers a fair chance of starting their new business success-fully. The consequence is that, first, political considerations must set the framework of licensing policy and the conditions which apply to the licences. This took some time and especially in Europe it had to be coordinated between the member countries of the EU. Even this process was difficult enough and was not finished by the time the licensing process of UMTS in the EU started. This caused many problems which would not have occurred if the process had already had the status as given in the latest Directive of the EU, 2002/21/EC 7 March 2002, establishing the regulatory framework and entering into force 25 July 2003 (Figure 6.1). But this is not the only framework and condition that influences licensing policy: the policy has to fit into the WTO rules as well. That can create many more problems, and one example is Decision No. 128/1999/EC of the European Parliament and of the Council on the coordinated introduction of a third-generation mobile and wireless communications system (UMTS) in the Community (European Commission, 1999). With that Decision the European Commission came into conflict with the US administration, namely US Trade Representative Charlene Barshefsky (Inside US Trade, 1999), who accused the EU of market protection, because the decision was in favour of only one 3G technology, UMTS. Eventually the European Commission

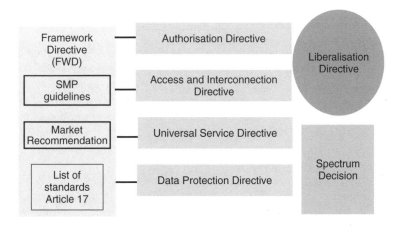

Figure 6.1 Structure of the EU Framework Directive 2002/21/EC. *Sources*: European Commission, ETSI.

under the leadership of Commissioner Bangemann was obliged to withdraw and modify the Decision. The outcome was that the Decision required that at least one UMTS licence has to be granted in each Member State to guarantee roaming between EU Member States – proof that politics and regulation are always interrelated!

But there are other reasons which require a regulatory process. One was mentioned in the EU Decision, namely roaming between networks and countries. Another very important one is spectrum. Spectrum is a limited resource, and therefore the most important part of a licence.

For each radio service you need a specific spectrum of frequencies. And worldwide frequency coordination is a very important task of ITU. Countries have agreed to accept and respect the frequency plan given and coordinated by ITU. So spectrum and radio service coordination is an issue for regulatory bodies.

6.1.3 SPECTRUM OWNERSHIP

As mentioned in the introduction, it is necessary to discuss this point in a little more depth. It is the mutual understanding in all societies that only governments should have the natural right to take care of scarce resources. In the case of frequencies this procedure guarantees, inter alia, the coordination beyond borders. Regulation of the right to use spectrum is essential to ensure that legitimate users obtain the required quantity and quality of radio communication services without disturbing interference. This is done by using licensing, frequency assignments, technical specifications and legal enforcement to ensure compliance. Without the appropriate use of licensing, including frequency assignments, the radio spectrum would become unusable.

There are two approaches to licence assignments:

- In the traditional approach licences are considered, especially as regards spectrum, to give the 'right to operate' or 'right of use'.
- A newer approach is closer to the concept of property transfer. In the case of spectrum, the State sells its spectrum to a private buyer, who can in turn resell it to other economic players. This leads to the idea of a market value attached to spectrum.

In the case of spectrum, the traditional approach is becoming less important and the newer approach seems to be based on the notion that spectrum is a property of the State that can be transferred to others. However, this approach may be incompatible with international commitments of the State, unless those international obligations are also transferred to the new owner(s). According to the ITU Radio Regulations, States are bound to comply with restrictions in the use of radio frequencies in their territories. Various limitations apply for frequencies for different applications, such as satellite communications, fixed services, broadcasting and mobile services.

In many countries, spectrum is now treated as land, which can be split up into parcels and sold or leased. The owner of a piece of land expects to use that land in any legal way that he chooses, as long as he does not interfere with his neighbours. However, spectrum is not so easily confined or delineated. The attributes and properties of spectrum are largely irreconcilable with the normal concept of property. This is especially clear in the case of spectrum rights in a border area, or spectrum rights to satellite frequencies.

Spectrum users must consider the use of spectrum in adjoining territories before they decide on their own use. No State can rightfully claim to have the full ownership of a satellite frequency, to the same extent as land.

Considerations such as these lead to the inevitable conclusion that spectrum is not 'owned' by any State. What the States own is the right to use spectrum in their own territory, within the restrictions given by the laws of nature, the treaty obligations in the Radio Regulations and the rightful interests of other countries.

6.1.3.1 International Regulation

The Constitution and Convention of the ITU together with the Radio Regulations constitute a treaty regarding telecommunications, including radio communications. Practically all countries in the world have joined the ITU and are bound by the Radio Regulations. These state that 'States shall endeavour to limit the number of frequencies and the spectrum used to the minimum essential to provide in a satisfactory manner the necessary services'.

In using frequency bands for radio services, states must bear in mind that radio frequencies are limited natural resources and that they must be used rationally, efficiently and economically so that countries may have equitable access to radio frequencies.

The ITU Constitution also lays down that all radio stations must be established and operated in such a manner as not to cause harmful interference to the radio services or communications of other countries.

Regulatory authorities coordinate and regulate their use of the radio spectrum with other countries on a global level within the framework of ITU at World Radiocommunication Conferences (WRC), which are held every 2–3 years, and within Europe through CEPT/ERC.

6.2 LICENSING

To better understand the difficult field of regulation and licensing, it may be helpful to explain the whole process by way of an example given by the UMTS Forum in its studies on 'The impact of licence cost levels on the UMTS business case' (UMTS Forum, 1998a) and 'Considerations of licensing conditions for UMTS network operations' (UMTS Forum, 1998b).

6.2.1 WHICH RIGHTS ARE TRANSFERRED IN A SPECTRUM LICENCE?

The price paid for a spectrum licence may, as stated in the previous section, be regarded as a rental or as an outright purchase of the State's user rights. The latter notion is logically combined with a right to resell the licence. Any transfer of rights would need to be registered with the spectrum management authorities and the spectrum market would need to be regulated to avoid abuses. Thus, under a market pricing mechanism the market itself would determine the price for traded spectrum rights.

Any discussion of spectrum pricing is likely to raise the question of 'privatisation'. Any kind of transfer or leasing of spectrum rights can itself be represented as a kind of privatisation but the logic can be taken further with the management of 'chunks' of the spectrum being put into private hands. The national authority then acts as a wholesaler, and possibly as a regulator of the kinds of intermediary which would retail spectrum rights to the final users of spectrum.

At present, most frequency licences throughout the world are non-transferable. If the right to transfer the use of sections of spectrum are sold off to private operators, it could hinder the development of international cooperation, as States may no longer control the use of spectrum and thus be unable to enter into international agreements. However, in no case known has a State completely parted with control of the licensed spectrum.

6.2.2 THE IMPACT OF LICENCE COSTS

The industry members of the UMTS Forum have studied the charges for granting licences, including spectrum licences, with special reference to UMTS, and have formulated recommendations from industry to national regulators concerning the use of charges. The regulatory members of the Forum have followed the study and contributed to it, but for understandable reasons have declined to be bound by the text and recommendations.

A charge generally accompanies the issue and use of a spectrum licence. Traditionally, these charges have been used as a means of financing licence administration costs. Another reason for the justification for pricing could be the encouragement of licensees to make the most efficient use of spectrum.

Since the number of potential licensees becomes smaller as the cost of spectrum increases, spectrum pricing can also be used as an instrument for selecting licensees in situations where there is a scarcity of frequencies.

Traditionally, the principle *First come, first served* is the most widespread and longstanding method of selecting licensees. *Comparative bidding*, sometimes called a beauty-contest, involves selecting the best applicant according to predefined selection criteria. *Auction*, sometimes called competitive bidding, awards the licence to the applicant that makes the highest bid.

Some studies undertaken by the UMTS Forum in 1998 suggested that where licensing fees exceed administrative costs there is a direct negative effect on the development of UMTS services. The results show that the profitability will be reduced and the payback period lengthened, and this deterioration is particularly marked when the licence fee comes to a level of about US$ 50 per capita. That has not changed much, as will be shown in Section 6.4.

If liquidity is a constraint for an operator, a licensing system based on annual fees may be preferable, since it allows the operator to spread payments over the period and escape large up-front payments. The lower the income for an operator, the greater the impact of licence fees.

For the mobile industry, high up-front licence charges increase investor uncertainty, and correspondingly decrease investment in the network for a new service such as UMTS. In addition, technology choices may be driven by a short-term focus on recovery of up-front fees rather than a long-term focus on overall growth of the industry. This may have potentially irreversible consequences for service provision.

High up-front licence charges will not facilitate the best services for UMTS consumers. They would increase the cost of services to consumers, and may price some consumers out of the market altogether. In the licensing and pricing of UMTS spectrum, governments seek to encourage innovative and efficient spectrum usage, and ensure the maximum economic return for the government. High up-front licensing charges will not achieve these goals. Up-front investments that are channelled not to spectrum access, but instead to extensive coverage, marketing programmes, high quality systems, robust networks and employee training will result in more effective use of the spectrum, and will in the long term result in more government income.

There were examples in the world which emphasised the possible consequences of disregarding these important factors. First, experience with the US PCS C-block auctions has shown how high charges can undermine a newcomer's viability when it has to roll out a totally new network starting from zero, while existing operators have the opportunity to reuse their infrastructure. This compromises competition. Second, two operators in Germany (QUAM and Mobilcom) pulled out of the UMTS market in 2003. They were not able to fulfil the moderate rollout requirements given in the German licence conditions. Bankruptcies and reduced investment incentives are the likely outcomes where high licence costs cannot be recovered quickly, because business is still in the developmental stage.

Probably because of the pressures of an auction situation, the market players decided to risk paying higher prices, although many sources, including the UMTS Forum, were warning of the possible adverse consequences.

To summarise, the cash flow generated from UMTS services is needed to cover both investments in the growing network as well as interest on the initial investments, which constitutes a heavy burden. The possibility of transferring costs to later stages in the business cycle is often a prerequisite for such companies. As shown with the experience of the German UMTS auction, a highly attractive market can lead to a situation where new entrants overestimate their ability to balance the funding of licence charges and investments in their network with their expected income. High up-front charges could in the short or medium term cause operators to go bankrupt. As very few operators may have the 'critical size' to maintain their activity on the UMTS market, competition could be seriously compromised.

However, looking at the long-term period of about 20 years, for which most of the licences were granted, eventually the paid licence fees may not play the dominant role for the economic success. With respect to the long-term period of a UMTS/3G licence operators may tolerate the high up-front payment of licence fees, but not all of them may run their network for the lifetime of the licence.

6.2.3 LICENCE FEES

Governments' spectrum management goals should include obtaining a fair value for spectrum licences in order to recover the costs associated with governments' spectrum management functions and/or to generally raise revenue. Internationally, ITU and ERC (on behalf of the European Union) are working on the issue of spectrum pricing mechanisms (CEPT/ERC, 1998; ITU-R, 1997), but it is very difficult to agree on an international basis on a legal pricing scheme or harmonized measurements.

The licence fee as developed includes a number of criteria, such as:

• frequency band
• bandwidth used
• exclusive or shared use
• geographical location
• coverage
• trunking efficiencies
• relative congestion.

In practice, a combination of criteria is used, in order to achieve a system that is fair, proportionate and transparent.

6.2.3.1 Use of Licensing Charges

In the USA, the Federal Communications Commission (FCC) is given the authority to collect regulatory fees, either for processing applications and/or for annual payments. The latter are used to cover the costs of the Commission for enforcement, policy and rulemaking, user information and international activities. The regulatory fees do not apply to government entities, amateur radio users and non-profit entities. The income from spectrum auctions in the USA is not used by the FCC but is delivered directly to the federal budget (Rosston and Steinberg, 1997).

Within the EU, the Licensing Directive (97/513/EC) (European Commission, 1997) provides the common framework used for issuing licences for telecommunications services. Individual countries in the EU have differing systems for collecting fees. In the UK and in Germany, for example, it works more or less the same way as in the USA. The auction prices far exceeded the governments' own estimations. In the UK the auction price was £22 billion instead of the £2 billion planned for the year 2000 national budget. This windfall received by the UMTS/3G licensing system went mainly to support the National Health Service. In Germany, the €50 billion received have almost completely been used to reduce the national debt. Only a small portion went into a national traffic improvement project after reunification.

Licensing is only one aspect of managing the spectrum, and it cannot function effectively without the support of other spectrum management activities such as overall frequency planning and coordination and monitoring.

The funding of all these activities can be based on either national budget financing or spectrum usage fees, or a combination of both. In addition to the issuing of licences there are other functions associated with spectrum management activities that may generate income, like type approval, test and certification, inspections and issue of certificates (radio amateurs, maritime examinations).

Providing the resources to perform all of the necessary spectrum management functions need not be confined to the administration. Some administrations are using private sector organisations to support specific spectrum management activities. For example, cellular operators normally undertake the frequency planning for their networks.

In most administrations, apart from auctions the way licence charges are set is strictly regulated. The fee may be applied to some or all radio spectrum users. There are currently three forms of spectrum usage fee in use, *simple fee*, *cost recovery* and *administrative pricing*. In practice, cost recovery might be considered as a variant of simple fee where the administration sets the value to cover the costs, but the distinction is made because its structure and operation are heavily influenced by national legislative and constitutional requirements.

6.2.3.2 Simple Fee

The simple fee may be set at the same level for all licences, or it may vary depending on the frequency band and service. The fee does not necessarily reflect the costs of the administration, so the fees recovered may be greater or less than the administration's costs.

6.2.3.3 Cost Recovery

The cost recovery charges for a radio frequency licence are set according to the costs for issuing the licence and for other necessary spectrum management functions. The exact definition and operation of cost recovery varies according to national spectrum management, legislative and constitutional requirements. These differences may have an impact on the implementation of cost recovery in each country and affect how the costs and fees are justified. Differences between administrations are particularly evident in the division between direct and indirect costs, and the types of costs permitted for inclusion as a basis for fee calculation.

In most European countries the licence holders of public and private radio systems contribute to the costs of spectrum management, although this may not be the case in those countries where the public network operator or broadcaster is a government entity. Where government spectrum users and broadcasters are charged, they do not usually contribute in a comparable way to other spectrum users for the spectrum they occupy. Indeed, in many countries government spectrum users do not make any payment for the spectrum they use.

Furthermore, the transmission capacity of broadcasters will be used increasingly for tele-communications purposes, and, conversely, telecommunication networks (e.g. UMTS networks) will be used for broadcasting or for relaying broadcast content to telecommunication subscribers. Since both types of networks will gradually convey overlapping content, they will compete in overlapping markets. Hence, they should be subject to equal financial treatment as regards frequency use, and there is a growing feeling that charge-free spectrum for broadcasters is unfair.

6.2.3.4 Administrative Pricing

Administrative pricing is a term often used for a charging policy under which the administration applies differentiated fees in order to influence the behaviour of spectrum users. In this approach, licence fees are set at levels that are not dependent on cost-based limitations. The objective of administrative pricing is to make the licensees use the spectrum more efficiently. It may also increase the supply of spectrum by making licensees:

* refrain from asking for more spectrum through use of more spectrally efficient equipment;
* hand back spectrum they do not need;
* move to a less congested part of the spectrum.

Administrative pricing may therefore also provide a mechanism to support a policy on spectrum refarming. Spectrum refarming means, e.g. revised allocation of spectrum used today for 2G which could later be used for 3G/UMTS as well. In fact, in Resolution COM5/25 of WRC-2000 the first- and second-generation mobile bands below 1 GHz (specifically in 806–960 MHz, e.g. GSM 900 and US cellular) are also identified, on a worldwide basis, for IMT-2000 to provide administrations with an opportunity to migrate these bands to IMT-2000 in the longer term (see Section 5.3.6).

In certain cases, administrative pricing may not be a necessary or appropriate tool to ensure the optimal use of spectrum because scarcity of spectrum and the highly competitive nature of the industry, among other things, already forces the players to use resources efficiently. The merit of administrative pricing mechanisms could also be diminished if only a few players bear the burden.

Administrative pricing cannot ensure that non-commercial users of spectrum adhere to the same principles of efficiency as the commercial users. Even if administrations use administrative pricing, they have to find means to make the non-commercial users adopt the most spectrum-efficient technology available and make appropriate investments in their systems to avoid waste of valuable spectrum resources.

6.2.4 THE VALUE OF SPECTRUM WHEN REUSED

A higher level of spectrum licensing fees can be motivated by a desire to raise the value and promote optimum use of spectrum. Income raised could be used to support the costs of refarming and relocation arising from changes of spectrum use. It could also be used for research and development of tools, e.g. software-based tools to improve spectrum management and efficiency.

The rate of spectrum refarming has increased substantially as private players intensify their use of spectrum and technologies change faster and faster. Traditionally, a spectrum user is forced by the authorities to move to other frequencies and invest in new equipment without special compensation. New players have seldom been asked to compensate the former spectrum users as the situation can be viewed as circumstances for which they are not responsible. This has led to the idea that the whole community of spectrum users bears the burden, with a refarming fund managed by the State, i.e. an additional charge levied by the State through spectrum fees.

Some countries require the new spectrum users to reimburse the old users of a particular frequency band in cases where the band is needed for new purposes. However, this may be decided on the merits of individual cases, and care must be taken that spectrum is released and relocation costs are levied in accordance with a reasonable profitability for the new spectrum users.

6.2.5 USE OF SPECTRUM PRICING AS AN INCENTIVE FOR EFFICIENT USE

The development of new wireless applications for some frequency bands in some geographical areas has resulted in a demand for spectrum that exceeds the available supply. Administrations have developed new economic methods to complement traditional methods of spectrum management. This leads to the notion of 'spectrum management by pricing' where the justification of pricing is also the optimisation of spectrum use. Spectrum pricing can thus, in principle, be considered as a tool for improving efficient use.

In theory, any level of spectrum charging that is not insignificant may be an incentive for the licensee, considering its business case, to limit its demand for spectrum and accordingly to use the spectrum more efficiently. However, the financial pressure of spectrum fees becomes an incentive only if an efficient use reduces financial pressure effectively.

High up-front fees do not push operators to obtain better efficiency, since whatever they do, the initial financial burden will remain unchanged. Such charges may, however, reduce the numbers of people using the services in this spectrum, because in order to recover the high up-front fees the operator may be forced to set high service tariffs. The net result could be that the spectrum is underused.

It is clear that the use of economic methods is not simply a matter of increasing the price charged for a licence, and does not guarantee that the spectrum is used efficiently. Additionally, the most appropriate spectrum for price-inelastic services, e.g. for social use, may become unavailable and the use of suboptimal spectrum may for technical reasons raise the cost of provision too high.

Optimal spectrum management cannot be achieved if financial pressure, in the form of annual charges, is not applied to the licensees until the end of their licence period. Mechanisms such as one-off payments that are made at the time the licence is issued (e.g. in many cases of auctions) could lead to a situation where operators 'freeze' their frequencies until the end of their licence, even if they stop operating many years before.

6.2.5.1 Telecommunication and Broadcast Services

The convergence between telecommunication and broadcasting services, as is anticipated today, will lead to a more intensive use of spectrum and should be accompanied by implementation of the same pricing principles. The introduction of third-generation systems, which provide an example of such convergence, should facilitate this development. It would distort competition and be unquestionably unfair if telecommunications carriers were subject to high spectrum fees, while others (e.g. broadcasting operators) were allowed to offer UMTS or UMTS-like services in existing spectrum without similar fee requirements. This disparity will increase as the convergence between telecommunication and broadcasting services takes place. In such a scenario, auctions are not competition-neutral.

6.2.6 USE OF SPECTRUM PRICING AS AN INSTRUMENT OF SELECTION

Since the number of potential licensees becomes smaller as the cost of spectrum increases, this can be used as an instrument for selecting licensees in situations where frequencies are scarce.

In discussing selection, it is important to distinguish between qualification criteria and selection criteria. Usually the administration sets criteria that form the entrance conditions applicants have to meet to take part in the selection. These criteria are called qualification criteria. Selection criteria are used to determine the winner of a licence.

6.2.6.1 First Come, First Served

Traditionally, the principle *first come, first served* is the most widespread and longstanding method. In principle, applications are dealt with in the order in which they came in. This approach is appropriate where there is no scarcity of frequencies. Private mobile radio (PMR) licences are a typical example. A large number of such spectrum users, who are not commercial operators and who in consequence do not compete among each other in the telecommunications service, obtain licences if the qualification criteria imposed by the NRA are met.

6.2.6.2 Comparative Bidding

Comparative bidding, sometimes called a beauty-contest, has as its goal the selection of the best applicant, according to predefined selection criteria. Economic factors are often among the qualification criteria, but such criteria are seldom exclusively used for selection. This method enables the pursuit of telecommunications policy goals and implies a thorough and systematic comparison of candidates. The downside is the risk of a complex and time-consuming procedure, subject to litigation.

6.2.6.3 Lottery

Lottery is seldom used in Europe, but has been used in the USA. With this method, all interested parties that fulfil the qualification criteria (if any), can apply for a part of the available spectrum. The award of the licence is not subject to a selection based on technical or economic criteria.

The advantage of the lottery approach is that, for the administration, it is a quick, simple and non-discriminatory method. The disadvantage is that there is no guarantee that it will result in choosing the most efficient operator, especially if the qualification criteria do not impose a high level of performance or there are difficulties in predetermining the relevant performance criteria, as may be the case with third-generation systems. Moreover, this method can lead to speculation if resale of the licence is allowed.

6.2.6.4 Auction

Auction, sometimes called competitive bidding, is a method where, among all candidates who fulfil the qualification criteria, the one making the highest bid is awarded the licence. The price of spectrum is determined by market expectations. Market theory predicts that auctions will allocate spectrum to those that value it the most and thus will make the most cost-efficient use of the spectrum. However, this does not necessarily mean that the spectrum will be used efficiently from the end-user's perspective.

Auctions resolve some of the problems identified with other procedures such as comparative bidding, if the auctions meet certain preconditions. They must, however, be prepared carefully in order to create a positive result, and this is time consuming and resource intensive. A survey of the characteristics of different auction types can be found in (Milgrom, 1989).

As in the case of a lottery, the effect of auctioning on frequency efficiency and development of the mobile market is also influenced by whether the licences are delivered on a transferable or a non-transferable basis and whether they are for a specified use or not.

One of the preconditions for auctions to function properly is that all potential players are fully informed. However, full information, in the context of UMTS, is not yet available, particularly with regard to spectrum availability, market demand and standards to be deployed.

As was explained earlier, a major risk with auctions is that the charges may become excessive. Auctions will probably increase the difference between what different players have to pay for spectrum for the same service, and what different services are valid for the same amount of spectrum. This will introduce market distortions. Auctions are therefore not a universal solution to all licensing problems and are not suitable for certain types of licences. For example, problems may occur when different applicants compete for a licence, e.g. broadcasting, scientific applications, civil defence use and other non-commercial uses.

When different radiocommunications services share the same frequency band, auctions are not very appropriate. Bidders' valuations will be discrete at a point in time. Such valuations may change due to new circumstances not predicted at the time of auction. It is hence vital that some mechanism is put in place to reassign spectrum.

6.2.7 USE OF SPECTRUM PRICING AS AN INSTRUMENT OF TAXATION

When the charging level becomes high, there may be more fees generated than can be spent on frequency-spectrum related matters. This amount should then be considered as a spectrum tax, not a charge.

Typical of a tax is that it is a monetary contribution to government funds without any direct service in return. A charge, on the other hand, is a monetary contribution paid for a specific service or performance by a government or a governmental agency. An imposed monetary contribution to government funds without a direct service in return can, however, also be considered as a charge, where the charge is used to fulfil a regulatory purpose with an indirect coupling to a service, such as spectrum management.

Using spectrum pricing for taxation may be more or less intended from the beginning. Some auctions have to a large extent been used as an instrument of taxation. Experience shows that once an element of taxation has been introduced, it is later hard to eliminate it.

The decision of the State to increase the charging level through spectrum pricing will have an impact on telecommunication market growth. Operators that are assigned spectrum will pass most of the costs on to the consumers and keep their profit level more or less constant.

Since it is well known that there is a close relationship between tariff levels and the growth of a wireless service, it is important for the State to consider to what extent the amount of taxation has a direct impact on the telecommunication market growth, and hence on the growth of the State's total economy. If charges are too high initially, it may damage the future base for value-added tax and other forms of income.

6.2.8 IMPACTS ON SOCIETY

Without question, high up-front licence fees are a cost that ultimately must be recovered from consumers through higher prices for services. Simply put, the higher the up-front charge, the higher the tariff that must be charged to recover that fee. And in the case of auctions, those up-front charges may be particularly inflated in order to win licence awards. The resulting high tariffs may, in turn, affect the consumer's take up of the service and ultimately, the success of the industry.

Up-front licence charges increase tariffs for consumers, slow the development of innovative services and therefore potentially harm competition. Rather than use up-front licence charges to allocate UMTS spectrum, other spectrum pricing mechanisms that encourage investment in the UMTS industry should be implemented. Such pricing mechanisms will be more beneficial to consumers and business interests, and therefore also to individual governments.

Auctions in one country will encourage auctions in another country, which could increase the risks that for economic reasons operators may chose an obviously 'cheaper' technology and that the world will become a patchwork of different standards and services in the same frequency bands, which in turn will significantly increase the difficulties of international coordination.

If many countries undertake such a selection process, the goal of global roaming will be defeated, as consumers will be able to use their terminals in some countries but not in others. Fortunately, operators in Europe declared in advance that they would go for the commonly agreed UMTS standard.

It must be realised that the value of a wireless telecommunications service is far more than the value of the spectrum it occupies. The dynamic effects on the economy are also important but harder to estimate.

A spectrum pricing policy that hurts the growth potential of the economy will prove to be very unprofitable in the long run.

6.2.9 CONSEQUENCES FOR THE INDUSTRY

As discussed above, when companies are forced to pay high up-front fees for a service with as-yet unproved market demand and experimental equipment, it can lead to business failures and bankruptcies – as was the case for Interactive Video Distribution Systems (IVDS) services in the USA or UMTS in Germany. The German experience with the UMTS licensing scheme provides a vivid case in point. As mentioned earlier, Quam and Mobilcom pulled out of the (UMTS) market. They did not reach the licensing requirement until end 2003. If they do not give the licence back to the NRA, they will be forced to do so according to the rules. In such circumstances, not only a government will be denied the tax revenues that are generated from successful businesses, but the government may also not even receive the up-front licence fee and end-user gets less competition as there are fewer service providers.

In addition, where service definitions have not been established and economies of scale in equipment markets have yet to be attained – as is the case for UMTS – auctions may not prove to be the revenue raising tool that some governments hope them to be. One example from the US auction experience – Wireless Communications Services (WCS) (UMTS Forum, 1998a) – proves that to be instructive.

Rather than look to high up-front fees to raise revenue, a government's interest in receiving appropriate compensation for public resources may be addressed through secondary benefits that are derived from the development of a new service – higher tax revenues as the service matures and broader economic growth from a more robust telecommunications sector.

Indeed, as the cases of IVDS and WCS spectrum auctions demonstrate, up-front fees may provide less economic rent from spectrum resources than otherwise would be obtained through faster spectrum exploitation. Because high up-front fees can increase prices and reduce demand, licences awarded using other criteria, without financial bids, may generate more taxes from operators and equipment suppliers, create more jobs and increase PSTN interconnection revenue. In the long run, the taxpayer will benefit more by the latter approach.

6.2.10 ASPECTS CONCERNING LICENCE CONDITIONS

6.2.10.1 Fixed–Mobile Convergence

In the UMTS environment the distinction between fixed and mobile networks will become increasingly blurred. If the full UMTS vision is to be realised within the envisaged time scales, the licensing of UMTS operators must allow for a convergent approach. Illustrative examples for this purpose are Deutsche Telekom and France Telecom.

In principle, there are no regulatory barriers to convergence today. Under their existing licences operators of mobile networks are usually not allowed to provide transmission capacity for purposes other than mobile services mentioned in the licence. However, they can apply for a new licence that would allow them to provide fixed services. Nevertheless, there could arise specific licensing problems (e.g. where fixed and mobile services and networks are subject to different licensing procedures).

The development of UMTS services should be market, not regulatory, driven. For this to occur effectively any regulatory or licence distortion, as well as artificial separation between market segments (e.g. between fixed and mobile) needs to be removed or avoided.

It also needs to be recognised that UMTS, as a service, does not start from a clean sheet. For that reason, migration paths from both fixed and mobile network infrastructures need to be enabled.

In the longer term the objective is to minimise regulation and to rely on competition law, while recognising that current telecommunications regulation is justified to establish effective competition.

6.2.10.2 Rollout and Coverage

Fundamental to efficient use of the spectrum is that spectrum is not left unused longer than necessary. Rollout and coverage obligations can be important incentives to achieve an efficient use of the spectrum and will encourage infrastructure competition, which in turn will lead to effective competition between operators and low prices for consumers.

On the other hand it must be recognised that the commercial pressures upon operators to roll out their networks are extremely strong drivers to achieve efficient use of the spectrum and optimise coverage levels to meet customer demand. Any rollout obligations should not, therefore, have a distorting effect on the UMTS market by requiring rollout levels beyond those that the market demands.

A negative example of rollout obligations, which caused a very negative impact, can be seen in countries like Sweden and Norway (Table 6.1). Operators asked the NRAs to adapt the obligation to the difficult market situation. As they decided not to change the licensing conditions, eventually some foreign operators (newcomers) pulled out, because they thought the rollout obligations were unreasonable with respect to the given time frame. The damage would not only be to the NRAs, but to the whole upcoming UMTS business in those countries as well.

Coverage obligations can be related to coverage of the territory (including coverage of the highways) or to a part of the population. As the geographical spreading of potential users may be very uneven, the best criterion would in most cases be based on a certain percentage of the population.

How much this would be and the time period in which the coverage obligation should be fulfilled may vary for different countries and markets. Different roads to achieve this can be followed, including national roaming and infrastructure sharing. It is important that regulators consider the constraints of technology when setting coverage obligations for terrestrial UMTS components.

Table 6.1 Deadlines for some UMTS rollout obligations

Countries	Month/Year											
	12/02	7/03	12/03	7/04	12/04	7/05	12/05	6/06	12/06	12/07	7/09	7/13
Austria			25%				50%					
Belgium		30%		40%		50%			85%			
France		25%									80%	
Germany			25%				50%					
Ireland					53%			80%				
Liechtenstein	70%		90%									
Norway					80%							99%
Portugal	20%						40%			60%		
Sweden				99.98%								
Switzerland	20%				50%							
UK										80%		

Source: IRG.

If an administration decides to impose a coverage obligation, it should clarify what is meant by coverage. In defining coverage one needs to take into consideration the data rate and the transmission power of the terminal, e.g. whether the coverage obligation should encompass:

- full wide area mobility (a bit rate of 384 kbps) with a handheld or
- only basic services with a lower bit rate or
- a portable terminal with more transmission power.

6.2.10.3 Roaming

International Roaming
Following the success of international roaming using GSM, international roaming between UMTS operators can only be encouraged. UMTS spectrum allocations should be in line throughout all regions, and the development of standards to allow international roaming should be encouraged.

National Roaming
Within the UMTS marketplace, consideration must be given both to roaming between the third-generation operators only, and to roaming between third- and second-generation operators. Effective long-term competition is best attained through a combination of infrastructure and service provision competition. This enables customers to benefit from lower prices, increased services and the ability to subscribe to services from an alternative operator. With a number of companies competing to meet the needs of the customer, innovation will be ensured, as it is essential if companies are to remain financially viable in the long run. Given this viewpoint, it is important that NRAs take account of the long-term competitive environment when developing their plans for licensing UMTS and in particular their approach to UMTS roaming. The impact of national roaming regulatory policies on infrastructure competition depends on

the state of network development in each country. For example, it is important that NRAs take due account of the costs and benefits of national roaming within the EU Member States to ensure that – whatever regulatory regime is put in place – it yields benefits to consumers, the UMTS industry and the economy.

National roaming can offer a consumer benefit allowing any customer to initiate or receive calls at locations which otherwise would not be served by their network operator. This could lower the barrier to market entry for new operators with no existing mobile access infrastructure and could provide for a better total coverage for the end-user.

This potentially could yield significant benefits for competition and customers. However, NRAs should recognise that significant benefits will only arise if efficient entry[1] is the norm. It is therefore important, if NRAs implement national roaming, that it is conducted in a manner that does not create an arbitrage opportunity where efficient entry is deterred while inefficient entry[2] is encouraged. Otherwise a distorted competitive market environment will be created to the detriment of the end-user, UMTS industry and economy.

If UMTS operators establish national roaming agreements from an early stage, it may discourage them from building their own networks, which could impact innovation in the marketplace at operator, service and content provider levels. With operators deploying their own infrastructure rather than using that of another operator, they can deploy leading-edge technology that facilities the development of advanced services. On the other hand, if incentives to invest in network infrastructure are low, technical and service innovation may not be assured.

Given the importance of infrastructure competition as outlined above and the conflict with national roaming, administrations must ensure that an appropriate balance is reached between national roaming and infrastructure competition.

This clearly depends on the state of network development in each country and could be countered by the appropriate use of rollout and coverage licence obligations, as well as by selective use (e.g. time-limited) of national roaming.

In the case of time-limited use of national roaming, care should be taken in setting the appropriate time period. Otherwise it is likely that a distorted market structure will be created, prompted by one operator's ability to 'free ride' on another operator's investments.

Currently customers who roam internationally accept service limitations when away from their home network and similarly there would be service limitations for national roaming customers. Customer education will promote the advantages relative to the limitations of national roaming. However, this may not be an issue if UMTS provides full service portability according to the virtual home environment (VHE) concept from the start.

Regulators should not lose sight of pricing, which is a key issue for any customer and could be of particular concern in remote areas. Given the substantial cost of building networks in remote areas, regulators must bear in mind that the ability to roam nationally could incur an additional cost as the cost of service provision may vary significantly.

---

[1] Efficient entry implies that operators are efficient in terms of long-run cost structures.

[2] Inefficient entry implies that operators are not efficient in the long run but can compete in the short run as a consequence of an arbitrage opportunity that enables them to exploit short-term profits at the expense of the long-term development of the industry.

6.2.10.4 Infrastructure Sharing/Facility Sharing

There might be significant economic reasons for the shared utilisation of existing or new network infrastructure. Sharing of infrastructure between network operators might help to reduce infrastructure cost and speed up network deployment. Consequently, infrastructure sharing might foster the rapid introduction of affordable services. On the other hand, infrastructure sharing might conflict with one of the primary regulatory goals: the introduction of effective infrastructure competition between public operators.

At a first glance, both goals may seem to be contradictory, though a precise understanding of the different levels of infrastructure sharing might be helpful to attain both goals.

A clear distinction should be made between:

- sharing of facilities (like sites and/or masts) and
- sharing of network infrastructure (radio, transmission).

In some European countries, the sharing of facilities like sites and/or masts among mobile operators is common practice. The arrangements are normally made on a commercial basis. Site sharing helps to reduce redundant cost in site acquisition, site development and mast construction. Moreover, it supports new entrants in raising competition as incumbents might have already occupied the best sites. For UMTS, site sharing will also reduce interference between networks and increase total capacity.

Site sharing is also important to overcome environmental concerns (e.g. protests against too many masts).

Hence, site sharing in the above-described way does not jeopardise the goal of infrastructure competition but may give rise to problems regarding property rights. On the other hand, the sharing of network infrastructure other than site sharing may cause conflicts with the idea of infrastructure competition. In that respect it is very similar to national roaming. Far-reaching network infrastructure sharing will definitely lead to the abolishment of infrastructure competition. The sharing of frequency and transmission capacities will result in negative effects on the willingness to compete on capacity and prices. Even the sharing of radio sites and transmission lines may endanger infrastructure competition, as competitors will get valuable information about the strategy of their counterparts. Moreover, sharing of network infrastructure might also lead to severe network integrity problems. However, allowing infrastructure sharing in rural areas would enable regulators to achieve a better total coverage of a country than when complete infrastructure competition is required, in the same way as has been pointed out before with regard to national roaming.

A good example is given in Germany and the UK, where with the blessings of the NRAs and the European Commission, O_2 (formerly BT Cellnet) and T-Mobile agreed on an infrastructure sharing concept, which gives both companies the possibility of increasing the market much faster for a better price, including many advantages for the consumer e.g. time-to-market products, earlier mobile broadband services and better price schemes.

In principle, to impose a mandatory sharing requirement on operators would entail undesirable interference in the market, but there might be specific cases where mandatory sharing would be desirable. Operators should retain discretion over the use of the infrastructure in which they have made substantial investment. Players should be allowed to negotiate among themselves in line with market forces, rather than having more barriers to competition with regulated tariffs.

In addition, each sharing situation is different. It would be practically impossible for the regulator to come up with a standard assessment of costs. For that reason, the UMTS Forum believes that sharing should be conducted on a commercial basis, with operators negotiating to set the appropriate level of tariffs.

Mandatory sharing would not be acceptable except as a means to fulfil the essential requirements of the Licensing Directive regarding environmental planning or in cases of limited access (e.g. bridges, tunnels).

6.2.11 REGIONAL LICENCES

Wide availability of service should be a key objective of any UMTS licensing policy. In this respect, lessons can be drawn from the success of second-generation mobile communications. While regional licences were appropriate and successful in some parts of the world (particularly in very large countries), most second-generation licences in Europe have been issued as national licences. Where regional licences have been issued in a competitive bidding environment, the operators have amalgamated several regional licences to form a national licence.

Regional licences, especially licences for small areas or even a single city, e.g. Greater London in the UK or the Rhine/Ruhr area in Germany, could lead to 'cherry-picking', resulting in certain areas receiving no service at all at the detriment of the end-user. Moreover, allocation of regional licences could lead to inefficient use of spectrum, as protection bands would need to be designated between different regions, otherwise spectrum interference could prevail.

These disadvantages combined with the success of national licences for second-generation mobile communications led, in the European case, to a clear vision in favour of national licences for third-generation mobile communications. In some other parts of the world regional licences (for large regions) could be a better choice. Two exceptions to this general rule can be envisaged. First, in special geographical circumstances (e.g. islands or remote areas), a regional licence could be used to cover an area, if it would not otherwise have had coverage from the national operators, e.g. the Isle of Man or the Channel Islands in Great Britain. Second, it may occur that an administration intends to license a total number of operators for which there is not enough spectrum available at the time of commercialisation. The administration in question may then consider dividing the available spectrum geographically between all UMTS operators. This means that temporarily, awaiting the availability of more spectrum enabling national coverage for all operators, the UMTS operators would only be able to operate in a limited geographical area, and the licence would have the effect of a regional licence during this period. This situation of temporary regional licences would have to be combined with a system of mandatory national roaming, to ensure that all subscribers of the several operators could make and receive calls nationwide.

It is important to ensure that this measure would not hamper the further availability of UMTS spectrum.

6.3 RECOMMENDATIONS FOR ADMINISTRATIONS

Those involved with the administration need to appreciate the developments in the mobile world during the past 5 years and take account of the lessons learned from the licensing and regulation process. The recommendations once formulated by the UMTS Forum seem to be still relevant and will suit best a successful licensing regime for a new mobile generation, especially when it comes to UMTS/3G. Most of them have been formulated by the UMTS Forum and can be found in their reports nos 3 and 4 (UMTS Forum, 1998a,b). Some extracts now follow.

6.3.1 NATURE OF LICENCE

Network licences for UMTS should be issued on a national basis to ensure wide availability of service. Special geographical circumstances or temporary national licences to be converted into a national licence when sufficient spectrum becomes available could justify an exception to this recommendation.

6.3.2 SELECTION OF LICENSEES

When selection of licensees is necessary because of lack of frequencies the administrative comparative approach should be preferred over auctions or lotteries. Auctions lead to high up-front fees, which will increase the tariffs for the consumers, slow down the development of new, innovative services, such as UMTS services, diminish the infrastructure investments and harm competition. Lotteries provide no assurance that a competent operator will be awarded a licence.

6.3.3 SPECTRUM PRICING

Spectrum pricing may be used as an incentive for efficient spectrum use, provided that these charges are fair, proportionate, transparent and competition neutral. They should mainly be motivated by cost-recovery and not by maximisation of revenue. The benefits, however, should be carefully weighed against the potential damage to the service.

Spectrum pricing as an instrument of taxation must be avoided, as it will have a direct negative impact on the growth of the telecommunications market and the general economy. Such taxation will in the long run diminish the total income for the State. High market values should be an incentive for regulators to find more spectrum, which will benefit the public more than excessive transfers of money to the public funds. Any regulatory actions regarding UMTS spectrum should be aimed at encouraging investments in UMTS systems. The UMTS business case study indicates that high fees will have a negative impact. Large downpayments at the beginning of the licence period should be avoided, in favour of charges related to the use of the system, like royalties or annual fees.

6.3.4 ROLLOUT OBLIGATIONS

Any obligations such as rollout and coverage should be appropriate to ensure competition in infrastructure and that frequencies are not left unused longer than necessary.

Service rollout of the networks should be based on market demands.

The burden of rollout obligations should be balanced between the operators' economic consideration and market demands.

6.3.5 ROAMING AGREEMENTS

Licence conditions should not prevent commercially negotiated roaming agreements to be reached between operators.

National roaming should not as a matter of course be mandatory due to detrimental effects upon competition and infrastructure rollout. It is recognised that under certain circumstances in a particular country mandatory national roaming may be justified as an exception. In these cases, the details of regulation should be carefully designed in order to minimise these detrimental effects.

Each administration may make its own decision on national roaming, taking into account their own state of infrastructure development and market competition and the objectives to be achieved. However, regional regulations (e.g. at EU level) should be taken into account, if doing so would be suitable to the market and, in particular, to the consumer.

6.3.6 FACILITY AND INFRASTRUCTURE SHARING

Facility sharing should be allowed on a commercial basis to foster the rapid deployment of networks and the introduction of services. Sharing of network infrastructure should be allowed, but restricted if it conflicts with the goal of infrastructure competition.

Mandatory sharing would not be acceptable except as a means to fulfil the essential requirements of the Licensing Directive regarding environmental planning or in cases of limited access (e.g. bridges, tunnels).

6.3.7 RE-LICENSING OF RETURNED SPECTRUM

However, regarding the scarcity of spectrum on a national basis there is another aspect to be discussed. In the case of bankruptcy of an operator, for example, NRAs should consider whether the available spectrum should be granted again. In general, that does seem quite reasonable. There are two options: either NRAs could start a new licensing process ruled under the existing licensing framework, where existing or new operators could apply, or, especially in countries where operators have paid a high up-front payment for the licence, NRAs could split the returned spectrum into (equal) portions and grant them free of charge to the remaining successful operators to guarantee the best service to customers. That would not increase end-user prices for mobile broadband services.

6.3.8 FIXED AND MOBILE SERVICES AND NETWORKS

Where fixed and mobile services and networks are subject to different licensing conditions, the administrations should clarify which licensing regime and conditions are applicable to converged fixed/mobile services and networks.

6.3.9 UMTS SATELLITE COMPONENT

Due consideration should be given to the particular global nature of satellite systems when seeking to establish a suitable mechanism for defining licensing costs and spectrum pricing for the UMTS satellite component.

6.4 LESSONS LEARNED FROM THE LICENSING PROCESS

As mentioned earlier, the recommendations given by the UMTS Forum have not been followed in all countries. The very first licence for UMTS was granted in March 1999 in Finland and followed almost word-by-word the UMTS Forum's recommendations (the top three of which were: four licences, each 2×15 MHz paired and 5 MHz unpaired, licence fees based just on administrative expenses). All Finnish operators and the NRA were members of the UMTS Forum and some were very active, so it was a great success for all parties, achieved quite smoothly and without any major controversy. This may have been because Finland is a smaller country situated in the north-east corner of Europe and not often in the international press spotlight. It may also have been because the NRA and the operators in Scandinavia had pioneered cellular mobile telephone services with the NMT system.

When it came to the UK and Germany to award licences, a very different situation developed because of two new requirements. First, NRAs decided to use an auction model, and, second, the administrations wanted at least one newcomer to enter the business. This made it more difficult for incumbents to get a suitable licence, as proposed in the UMTS Forum's studies and recommendations. A very competitive situation developed in which the incumbents feared that there would be no future to their present business if they failed in the licensing process. The intense competition resulted in overbidding for a licence which ultimately did not fulfil their expectations. On the other hand, you also have to see that the desperation of existing operators not to be left out of the new revolution was partly fuelled by the apparently unlimited success of 2G and the dotcom fever in general. In other words, there was a fever in the market which has now gone, probably never to return.

In the UK, among the field of incumbents, only Vodafone with a very clever bidding strategy was awarded a license with sufficient bandwidth (2×15 MHz). '3' as a newcomer obtained the same spectrum, but for less money. However, 3's disadvantage was that it had no 2G network with its customer base. Not only were there no revenues to help fund the investments required to build the new network, but there is no customer base to migrate to the new network.

Germany was different to the UK as all bidders had the same chance to obtain a licence of n-times 5 MHz spectrum. There was no separate bidding for newcomers and incumbents and all fought on equal terms. When the phase of the licensing process was over, the first bidders pulled out and already some €30 billion were committed to the government. Most observers expected that the six remaining bidders would share the spectrum, leaving 2×10 MHz for each of them. But the incumbents tried to pressurise the remaining two newcomers out with the hope of receiving 2×15 MHz each which would perfectly serve their needs. Surprisingly, the two newcomers (Group 3G, later Quam – supported by Sonera and Telefonica – and Mobilcom – supported by France Telecom) continued in the bidding process, which ended at a total cost of €50 billion. The question of who benefitted most from this auction process is still to be decided: the two newcomers surrendered 3 years later; the incumbents found they could not tolerate the high expenses and started internal cost-saving programmes. The taxpayer will not see any tax revenues from these companies for many years.

What was the effect of these experiences on other countries? On the whole, the operators paid much less in the licensing processes in other countries. Even where the licence fees remained quite high, like in France, some global or regional players did not show any interest and consequently, some administrations lowered the fees or gave the operators some additional benefits like extended lifetime of the licence. The financial implications for the

Table 6.2   Allocation of 3G mobile licences in selected economies worldwide

Country	Number of licences	Mobile incumbents	Method	Date awarded	Amount paid (US$ million)
Australia	6	3	Regional auction	March 2001	610
Austria	6	4	Auction	November 2000	618
Belgium	4	3	Auction	March 2001	421
Czech Republic	2	2	Auction	December 2001	200
Denmark	4	3	Sealed-bid auction	September 2001	472
Finland	4	3	Beauty contest	March 1999	Nominal
France	4 (1 pending)	3	Beauty contest + fee	July 2001/ September 2002	4 520 (subsequently reduced to 553 million each, plus 1% of revenue
Germany	6	4	Auction	August 2000	46 140
Greece	3	3	Hybrid	July 2001	414
Hong Kong, China	4	6	Hybrid	September 2001	Minimum 170 each plus royalties
Israel	3	3	Beauty contest + fee	December 2001	157
Italy	5	4	Hybrid	October 2000	10 180
Japan	3	3	Beauty contest	June 2000	Free
Korea (Rep.)	3	2	Beauty contest + fee	August 2001	2 886
Malaysia	3	3	Beauty contest	December 2001	Nominal
Netherlands	5	5	Auction	July 2000	2 500
New Zealand	4	2	Auction	January 2001	60
Norway	4	2	Beauty contest + fee	November 2000	88
Singapore	3 (+1?)	3	Cancelled auction	April 2001	165.8
Slovenia	1	2	Cancelled auction	December 2001	82
Spain	4	3	Beauty contest + fee	March 2000	480
Sweden	4	3	Beauty contest	December 2000	44
Switzerland	4	2	Auction	December 2000	120
Taiwan, China	5	4	Auction	February 2002	1 400
United Kingdom	5	4	Auction	April 2000	35 400
Total (25)	99+	79	13 auctions, 9 beauty contests, 3 hybrid		105 330+

Sources: ITU, European Commission, UMTS Forum and 3GNewsroom.com.

operators resulting from the auctions in the UK and Germany caused a very negative reaction in the stock markets. The share prices of both operators and their hardware and software vendors dropped by almost 90%. There were other things happening in the global economy at the same time. Finally, you cannot blame the auctions alone for the drop of share prices in 2000. Now, in 2005, the situation seems to be consolidating, as operators are better able to convince financial analysts that their services and market strategy are more realistic. These strategies rely more on an evolutionary path from existing 2G with a combination of WLAN and fixed network services. However, the situation for newcomers remains difficult. As about 115 licences are granted to date (end 2004) an incomplete summary of 3G mobile licence allocation in selected economies worldwide is shown in Table 6.2 (ITU, 2003).

As more countries pursued their own variations of licensing, the UMTS Forum (Sidenbladh, 2002) analysed the licensing regimes and results which occurred in the first round. The analysis is shown in Figures 6.2–6.4.

1. The licence fees differ very much between almost zero (e.g. Finland) and €180 per head of population for 2×15 MHz in the UK. Even more was paid (€360 per head of population and 2×15 MHz) in the USA in the New York area in January 2001 when the so-called Nexwave spectrum was re-auctioned. That spectrum within the 1.9 GHz band can be used for 2G and 3G as well, which made it interesting to operators which were already offering 2G services in that band. Figure 6.2 illustrates that impressively.
2. It makes a big difference whether you apply for a licence in a big or a small country and whether the licensing process is governed by beauty contest or auction. Figure 6.3 shows that global players all in all have the best chances, especially in auctions. That should give NRAs a clue when they establish a licence process and operators an argument when negotiating the licensing regime with NRAs.

Another quite interesting analysis of 2001 is shown by Booz, Allen & Hamilton (Figure 6.4). It says that only operators that were prepared to spend at least €15 billion on licences would have been able to cover more than 150 million of the population in Western Europe. This means that only big operators can play a significant role regionally or globally.

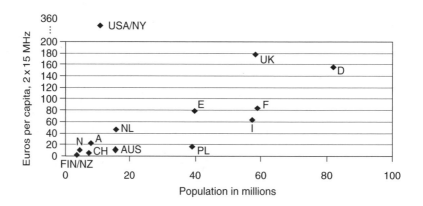

Figure 6.2   License fees in €/capita/2×15 MHz. *Source*: UMTS Forum.

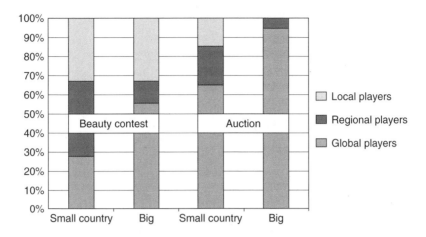

Figure 6.3 Probability of winning a licence and operator categories. *Source*: UMTS Forum.

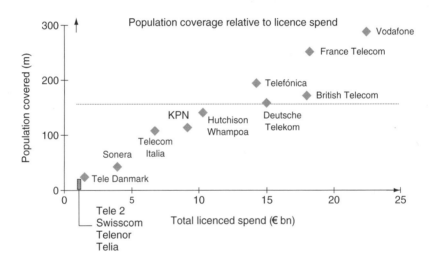

Figure 6.4 Only spending of more than €15 billion ensured operators a population coverage of more than 150 million in Western Europe. *Source*: Booz, Allen & Hamilton, 2001.

Those operators could cover a significant population and, in the light of consumers, only they would be able to offer regional/global services and, eventually, to operate on a global brand.

6.5 GLOBAL TERMINAL CIRCULATION AND CERTIFICATION

Global terminal circulation means the right of a user to carry and use his/her mobile terminal on any compatible network in any visited country, subject of course to suitable roaming agreements and commercial terms. Global terminal circulation is user oriented. In contrast, global terminal certification is product and manufacturer oriented and means the right to sell mobile handsets worldwide.

Apart from any necessary regulatory approvals, global terminal certification needs a Conformity Assessment Agreement (CAA) within the global WTO framework, based on:

- Supplier's Declaration of Conformity (SDoC) – without mandatory third-party involvement;
- market surveillance by national authorities;
- no national marks on the product.

To facilitate the achievement of Global Terminal Circulation, the Regulatory Aspects Group (RAG) of the UMTS Forum issued in October 1999 a so-called High-Level Statement stating:

- personal terminals should be exempt from customs duties;
- no individual licences should be required for personal use;
- terminal design must avoid unacceptable interference;
- terminals must comply with unwanted emission limits.

In addition, the Trans-Atlantic Business Dialogue (TABD), a group comprising representatives of the European Commission, some governments of the EU, the government of the USA, and supported by industry experts from both sides of the Atlantic commissioned a Communiqué in Berlin, November 1999, calling on governments to remove global circulation barriers.

During the past few years, the ITU has studied global terminal circulation rules and conditions, and via its Working Party 8F ITU submitted the following Recommendations:

- IMT.UNWANT-MS & BS concerning emissions from terminals and base stations;
- IMT.RCIRC concerning principles for global circulation.

Significant accord has been reached within the ITU that finally clears the way towards an effective and practical blueprint for global circulation. In October 2001, a major milestone was reached at a meeting in Tokyo hosted by ITU-R Working Party 8F, where governments all over the world, supported by standards development organisations and other industry bodies – including the UMTS Forum – agreed details of a regulatory framework ensuring that people can travel with their 3G terminal and use it anywhere in the world, free of regulatory and administrative hurdles. Building on this decision, a special ITU task force continues to discuss the implementation of global circulation with political bodies at the highest level.

On a European level, CEPT/ERC and ERO have issued Report 60 'Global circulation of IMT-2000 terminals' (CEPT/ERC, 1998), which contains the following sentence: 'Administrations should work together to develop a policy document . . . as a first step towards the facilitation of global circulation for IMT-2000 terminal equipment'.

Also in Europe the Radio and Telecommunications Terminal Equipment (RTTE) Directive, which was introduced to open up the markets for radio and non-radio terminal equipment by ending the national type approval regimes, calls for:

- Minimal Regulatory Approval technical requirements;
- focus on essential health and safety, EMC and effective spectrum use;

- open systems interfaces (OSI) to increase competition in terminals;
- preference for harmonised standards for interoperability.

Also Customs authorities have prepared the path for global terminal circulation: personal terminals are to be treated as personal effects under the Istanbul Convention on Customs Duties (World Customs Organization, 1990; ITU, 2002; 3GPP, 2000).

Returning to the UMTS Forum activities in this field, the subject was on the agenda of the 3GSM Word Congress in 2002 in Cannes. Here the UMTS Forum chairman's statement on global terminal circulation published in the 3GSM Show Daily contained the following remark (Eylert, 2002):

> Less glamorous than shiny new handsets and talk of exciting services, global circulation of mobile terminals nevertheless remains crucial to successful market uptake of 3G. With confirmation from the ITU of a technical framework that will streamline the circulation of IMT-2000 terminals, 3G's future as a truly global system is ensured.

On the face of it, it may seem obvious that mobile users can take their handsets with them and use them wherever they want to in the world – network availability and roaming agreements between operators permitting. With GSM, international roaming has played a major part in ensuring the standard's incredible success. Indeed, roaming is such a central part of today's GSM customer experience that it is difficult to remember a time when you wouldn't automatically slip your phone into a pocket when travelling overseas. With GSM, this ability grew out of a single standard with three frequency bands with operators controlling circulation and encouraging international coverage through roaming agreements.

Similarly, in the context of UMTS, global circulation can be defined as 'the ability for users to carry their IMT-2000 terminal with them anywhere in the world and to use it wherever transmission is authorised'. This is a major part of the 3G vision of ubiquitous services that can be accessed anywhere and any time, and users of IMT-2000 terminals will expect no less than their experiences with GSM. In addition to the obvious utility that global circulation offers end-users – as already demonstrated with the strong market appeal of roaming with GSM – global circulation will generate additional revenue for operators, as well as helping countries take part in the growing market for mobile multimedia that 3G promises. Unhindered global circulation is thus an important issue for governments and UMTS operators, but what are the specific steps that our industry is taking to make it a reality?

As the mobile component of tomorrow's information society, 3G will mean an explosion in terminal device types, form factors and functionalities to support an enormous range of services and applications. With product development cycles getting ever shorter, 3G will herald a new generation of personal information devices that will blur the boundaries of 'traditional' telephony and Internet access. There are those who suggest that network performance may be downgraded with a flood of different terminals from many manufacturers – they fear that models that are not formally approved for use with a particular network may exceed prescribed emission limits and disturb other radio traffic. In 3G, new players and new countries will be joining the roaming experience, and some of these have claimed that new regulation is needed in order to 'guarantee' the performance and integrity of their new offerings. But as you have already learned from GSM, with 3G it will be the commercial interests of all industry players that provide the market with all the regulation it needs.

Whereas GSM grew to become a de facto standard by organic adoption by individual countries inside and outside Europe, with 3G the picture is subtly different. First and foremost, IMT-2000 is in its very essence a global project, and as such is being introduced more or less simultaneously in many countries that lie outside the GSM footprint. Second, IMT-2000 gives the operators several technological options and will deliver a service portfolio increasingly more complex than that which GSM can deliver. This means that the technical implications of the circulation of 3G terminals could be more complex than with GSM: facing the complexity of IMT-2000's global deployment, a first reaction might be to introduce stringent regulations that inhibit the carriage of 3G terminals to protect network operators and end-users. This response, however, is inappropriate and does not address the core of the problem, but instead threatens to stifle innovation and natural market growth. It is important that the chosen regulatory approach does not imply extra administrative burdens when compared with technologies such as GSM or WLAN. With undue regulation comes the risk of damaging the market opportunity for new, exciting products and services that 3G will stimulate.

In the light of the industry, avoidance of the creation of new regulations in areas where commercial players can handle the issues by themselves is paramount. In the case of global circulation, 'visiting' terminals will not create any new problems that may not already have been experienced with 'local' terminals. Many UMTS/IMT-2000 terminals will include a number of operational modes, adapted to various frequency bands and air interfaces that are not be supported in all countries. It is crucial that the various modes of the terminals do not give rise to unwanted RF interference when it is switched on in other countries. A simple solution to this is adoption of the receive-before-transmit principle, ensuring that the handset does not 'chatter' until it has received a validation signal from the network that it is visiting. Each operator must solve problems of interference in their network when designing and deploying the network. The operators will then chose roaming partners among other operators whose networks are performing well. Against this background it is reasonable to expect that the operators themselves can handle all problems relating to their own terminals as well as visiting terminals.

In conclusion, let us quote the chairman's final remark at the above-mentioned event to look at the marketing impetus as well:

"...governments now have an effective 'toolkit' of guidelines courtesy of the ITU that can help the smooth realisation of 3G global terminal circulation. Operators don't want – or need – the imposition of onerous regulatory straightjackets: indeed, it is they who are taking the commercial risk, and as such they have quite enough incentive to provide their customers with an excellent roaming experience. By the same token, it is the manufacturing community who stands to lose most from any substandard performance of their terminal products. Building on the enormous success of 2G, all players in the 3G market are poised to deploy systems and services that offer end-users seamless access to an exciting new world of mobile multimedia...one that's even now becoming a reality."

6.5.1 CONCLUSION

Key principals for global circulation of UMTS/IMT-2000 Terminals are as follows:

- The circulation of UMTS/IMT-2000 terminals intended for personal use should be exempt from all customs duties or other official charges.
- The personal use of UMTS/IMT-2000 terminals should require no individual licence or any other form of individual formal regulatory procedure.
- Terminals shall not cause unacceptable interference in any country where they circulate. One way of achieving this is the application of the receive-before-transmit principle.
- Terminals shall comply with unwanted emission limits.
- Authorities should cooperate in order to enable global circulation of such terminals in all parts of the world.

With these key principals, global circulation of UMTS/IMT-2000 terminals will be successful for the benefit of the end-user.

7

Mobile Communication and its Impact on Social Behaviour

Note: This chapter is a copy of an article conducted as a summary of the report 'Social shaping of UMTS – preparing the 3G customer' (UMTS Forum, 2003a) by the Digital World Research Centre (DWRC) on behalf of the UMTS Forum for the *European Information Technology Observatory 2003* of EITO. The UMTS Forum, under the chairmanship of the author and the guidance of its programme manager Steve Hearnden, produced this article. The author is delighted to express his big thanks that EITO has generously given its explicit permission to reproduce it for this book. DWRC is an institute at the University of Surrey, UK. At the time when the study was conducted its Director was Professor Richard Harper. The research fellow was Jane Vincent. EITO is an initiative of the communication and information industry in Europe which produces an annual yearbook on the same subjects. In 2003 DWRC undertook a subsequent study, 'Informing suppliers of user behaviour to better prepare them for the 3G/UMTS customer', on this item for the UMTS Forum, published as UMTS Forum report no. 34 (2003C) and available on the website at www.umts-forum.org.

7.1 EXECUTIVE SUMMARY

The research programme has identified and explored some of the key ways in which consumer needs and expectations have driven GSM technology and how those same or similar needs will shape future UMTS services and products. The purpose of this is not merely scientific, but is intended to enable suppliers of UMTS products to be in a better position to plan, shape and develop their offerings. It will also enable them to develop more effective strategies for shaping customer expectations.

The research was conducted in three phases.

The aims of the first phase were to identify key social drivers for the development and success of GSM that may be applicable to the introduction of UMTS technology. The results of this phase of the Study are shown in Section 7.5 of this chapter.

The second involved deepening understanding of a subset of those themes, selected in close cooperation with the UMTS Forum working party on the project. Further data gathering and research was then undertaken, before initial drafts of the implications that these social shaping drivers would have for UMTS services and products were prepared.

Finally, phase three involved presenting those implications to key players in the industry, to test for their accuracy and relevance, and to enable a further refinement so as to deliver analysis of the implications that combine scientific and commercial–industrial understanding of what social shaping will mean for UMTS.

The key implications for suppliers of UMTS relate to the three themes investigated in depth. Each theme was supported by a hypothesis that was used as the basis for the more detailed exploration using largely qualitative research and state of art literature review techniques.

These themes and the implications for each are:

1. Mobile devices do not enable more social relations but in contrast more intensive relations with already existing contacts.

The key implication of this for UMTS is that GSM technology has been shaped to satisfy the need for 'personal telephony' rather than 'mobile telephony'. This will mean that it will be difficult, though not impossible, to extend users' requirements for UMTS products and services into non-personal needs, especially business needs. Nonetheless, there will be significant opportunity for expanding and enriching the experience of personal communications (such as through imaging).

2. Users have a more emotional relationship with their mobile phone than they do with other forms of computational device.

The key implication deriving from this is that UMTS services and products that satisfy emotional needs will consist of person to person connectivity applications. The social value of these services will be much higher than the value given to person to information services. This will be reflected in the price sensitivity of each genre of application.

3. The intersection of public by private behaviours enabled by mobile phones will reach a threshold beyond which resistance will start to occur.

The key implications from this are that, in the case of Location Based Services (LBS), user and regulatory resistance will occur if these types of applications are introduced without enabling more fine grained permission-based control than is currently available. This will require considerable improvement in the man–machine interface (MMI) of UMTS devices that will ensure that permissions can be provided on an ad hoc, 'need to know' basis.

The implications for imaging applications are that they will find wide acceptance if they encourage users to develop a form of use that is analogous to texting. This will involve such things as 'pictures for play', for example. There is a strong likelihood that a considerable market will emerge for imaging, if the shaping of consumer expectations can be done effectively. However, the implications for video telephony are that it will generate considerable resistance unless radical improvements in the MMI and the form factor of handheld devices is achieved which allow much more flexible management of the social etiquette. Currently, these changes are not on the roadmap for terminal design.

7.2 THEME 1: SOCIAL CONNECTIVITY

The theme had to do with the assertion that mobile phones are increasing connectivity, which is not manifested in calling more people, rather, it increases the frequency of the contacts between the same people. Moreover, a further related suggestion was that mobile phones are creating an opportunity for new and informal relationships in business. If this is true then what was being sought in this tranche of research were the implications this has for 3G.

For this inquiry the Study used data obtained from a specially commissioned questionnaire that was circulated in the UK, France and Germany. This was combined with other data from focus groups and ongoing research at the DWRC.

To summarise, the Study confirmed that it is the previously existing social groups of people who are calling more often, and that there is higher frequency of calling. This is essentially for social purposes: friends and family keeping up with the action or business colleagues keeping in touch. The Study also found that people are unlikely to say they use their phones for fun and to alleviate boredom because of cost sensitivities to these activities.

7.2.1 FINDINGS

More particularly, the primary value of the mobile phone is perceived to be for functional activities, especially for talk and text but not, apparently, for fun. However, the evidence also confirmed that the use of mobile phones strengthens existing relations but does not widen social connectivity. Relationships with business clients over the mobile are minimal and mostly talk; perhaps more importantly, mobile phones are used less for communication with business clients than with business colleagues, family and friends (Figure 7.1).

Though there was a keenness to avoid suggesting that this happened often, respondents also indicated that they mostly use talk with family in the same household and mostly use text with close friends. Mobiles are used more for staying in touch with business colleagues than with business clients (Figure 7.2).

Though the topics for this theme of the research had to do with social connectivity, certain issues related to the fears that go with mobile phones also raised themselves and these are worth mentioning. Many subjects expressed a concern that health issues were still unresolved, for example, and this led some people to lower their use of mobile phones. More importantly for the research theme, however, was the feeling that many people had that the mobile phone is increasingly becoming the only place that they have all their social and family phone numbers. Consequently, the potential loss of these numbers could create havoc and thus people worry about it happening.

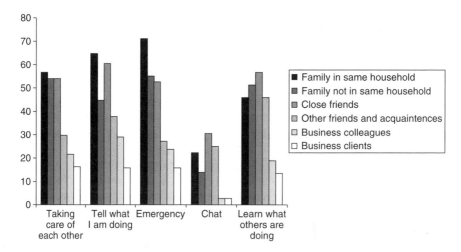

Figure 7.1   Percentage of respondents who agree they use their mobile phones for these types of communications. *Source*: DWRC Questionnaire 2002.

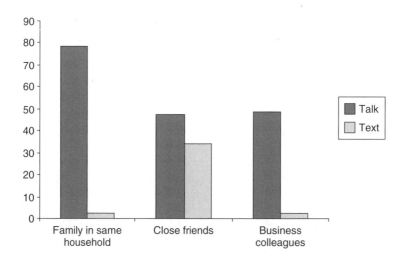

Figure 7.2   Percentage of respondents who agree that they mostly use talk and text with family in same household, close friends and business colleagues. *Source*: DWRC Questionnaire 2002.

7.2.2 *INTEGRATION WITH OTHER DATA*

When considered in relation to other data, there was a confirmation that people use mobile phones to achieve what one might call the functional prerequisites of social connectivity – organising their lives and keeping up 'with the action' (as they say). Yet across the board (and not only in the survey data) people do not admit to using mobile phones for what one might describe as non-functional, low perceived value activities such as for fun and to alleviate boredom: in other words, the perceived utility value of a mobile phone is to be in touch with close friends and family, it is not viewed in terms of fun, or more generally for person to information interaction.

7.2.3 IMPLICATIONS FOR UMTS SUPPLIERS

The key lesson from this is that people are not expanding their social horizon with mobile telephony. Rather, they are using the technology to organise their life and preserve (and indeed deepen) their social connectivity. The value of this social connectivity is great, and so there may be some cost inelasticity in the use of the mobile to support it. In contrast, the use of the mobile for fun or to alleviate boredom may be creating costs that cannot be justified by the benefits accrued. Indeed it could even be said that such activities may be a step too far if costs become too noticeable.

Given this, the primary implication for UMTS suppliers is that they need to assume that their products and services will be taken up by members of already existing social networks (consisting of personal and business contacts) rather than through entirely new networks of users, such as a new breed of post-3G mobile professionals, for example.

If this is so, then UMTS suppliers can map out these networks and use various tools to market to them, in much the same way that fixed line operators have offered 'friends and family' deals for many years. In this way, UMTS suppliers would be reinforcing the behaviours that have generated the demand for GSM networks. Having built on these social networks, UMTS might then be able to develop opportunities to start encouraging other uses of the UMTS services, such as for M-commerce.

However, this may be difficult to do. This is because there may be substantially different cost elasticities for social connectivity services as against other types of services, such as M-commerce. These two types of services may be called person-to-person services and person to information services, respectively. The latter sets of services may be thought of as analogous to those transaction services provided on the WWW. The evidence shows that people are willing to pay a healthy price for their person to person services (since it sustains their social lives, and thus has a high social value) but that they might be much more sensitive about the costs of information use. Given that access to information in the broadest sense is a key goal of UMTS, this needs to be addressed very carefully indeed.

7.2.4 INTERVIEWS WITH EXPERTS: REVISIONS OF THE
IMPLICATIONS

The expert panel comprised people from a range of countries, operators and terminal manufacturers. Despite this wide spectrum of backgrounds and perspectives, there was consensus that the preceding analysis was correct. All agreed that GSM devices are used for social connectivity. It is in light of common understanding that the UMTS industry is developing new product propositions designed to address social connectivity needs.

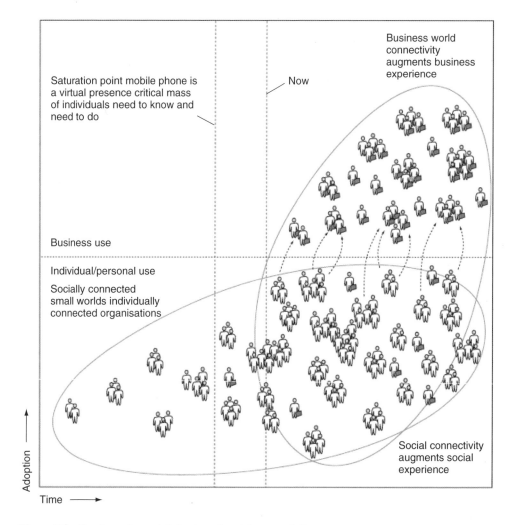

Figure 7.3   Roadmap for social shaping of social connectivity.

Explanation

This figure shows that the adoption of mobile devices sustains fairly small social worlds. As saturation levels increase, so the numbers of the social worlds supported increases too. Only a small proportion of these social worlds consist of business to business use. In the longer term, the intensity of the social connectivity communications may increase and be enriched by more services and applications. Costs for these services will be inelastic. In the longer term, there may be an opportunity for augmenting social connectivity among business users.

In the shorter term these products have to do with such things as sharing messages and images that relate to the concerns of the particular social group in question (such as 'football goal of the day' etc.). In the longer term, the expert panel believed that the UMTS market can be expanded by using those social connectivity applications that have succeeded on the Internet and delivering versions of these over UMTS networks. Here, mobile email, instant messaging, and Internet chat rooms were all cited as genres of social connectivity that might help expand the UMTS marketplace. Underscoring this vision was the belief that UMTS products would complement and thus expand those offered over fixed line Internet. This would encourage greater adoption of mixed media among users, something that the panel believed acceptable given the take-up of SMS by consumers.

There were some differences between the panel and the research analysis. One minor difference had to do with the belief that quality of service over mobile networks is a greater inhibitor than the research evidence suggests. Such a difference would be expected to disappear with UMTS, of course.

More importantly, only a proportion of the panel believed that social connectivity services would be more price inelastic than person to information services, though those who did recognise this also believed that there would be significant ways of leveraging these differences (Figure 7.3). One suggestion was to use different price mechanisms for person to person connectivity as against person to information connectivity.

7.3 THEME 2: EMOTION AND MOBILES

Research in phase 1 produced the hypothesis that users have stronger 'emotional' relationships with their mobiles than they do with other forms of ICT. This was meant not in the sense that they covet the device (as in a symbol of a life style) but because they have come to depend upon it in ways that is crucial to their emotional lives. The Study sought to ask whether this is true and if so, what implications does it have for 3G?

For this, the Study combined state of the art research plus newly gathered first hand evidence. This evidence was generated by focus groups undertaken in the UK and Germany. Data from these were to used to explore the hypothesis in terms of how people 'account' for their emotional experience with their mobile phones and to distinguish between sayings and actions. Thereafter, some further analysis was undertaken by looking at the use of the term emotion when exploring relationships with other forms of technology. This was integrated before implications for UMTS were defined.

7.3.1 FINDINGS

Few people use the term emotion to describe their relationship with mobiles: 'It's a funny way of putting it' being a common response to the proposition. However, most people do in fact use emotional language categories to explain their mobile usage: these categories include panic, need, desire, anxiety, etc. Users also achieve emotional goals with their mobiles as well as undertake emotional behaviour. For these reasons, the claim that there is a highly charged emotional relationship with mobile devices is correct.

More particularly, the emotional language categories used to account for the mobile experience are listed in Table 7.1.

People's use of mobiles for emotional goals has a number of facets. Perhaps the most obvious has to do with setting up social arrangements: 'I call my friends...stupid calls...I'm meeting them in half an hour and I'll call them, speak to them...until I meet them'.

Another has to do with avoiding making set appointment times – just arrange to call when you get there: 'meeting in a big park of people'.

Table 7.1   Summary of concerns about emotion and mobile phones

Emotion	Explanation
Panic	Absence from the device; being separated from it
Strangeness	Between those who do and those who don't have mobile phones
'Being cool'	Chilled out, tuned in to the mobile phone culture
Irrational behaviour	Can't control heart over mind, e.g. driving and talking
Thrill	Novelty, multi-tasking, intimacy of the text received in public
Anxiety	Fear and desire: e.g. not knowing and wanting to know about others versus too much knowledge

And a third (though there are more) has to do with making or breaking relationships: 'You can be silly on texting, you're too embarrassed to phone'; 'If I want to speak to my girlfriend any time of the day I know that I can and it kind of takes the fun out of it when I'm seeing her'.

In addition to emotional goals, users also behave 'emotionally' in the sense of behaviour in irrational ways. They constantly call their partner/spouse, for example, even when they are in the same house: 'I just feel the need to'.

They use the mobile impetuously: 'I just had to call someone'.

And even though they know they should not, they use it in places that creates danger: 'Even when I am driving and I go over those mini roundabouts in 4th [gear]'.

If these are three rather closely related issues, there was also a larger view. Users think the mobile phone helps them enjoy their life more: 'But it's not changed who I am'. They think that the mobiles are important: 'But might be getting too dependent on them' and lastly, the importance is so great it makes the mobile too valuable: 'I don't take it to the club 'cause it would be terrible if I lost it'.

7.3.2 INTEGRATION WITH OTHER DATA

General research on emotion mostly focuses on how form and function create emotional desire for a mobile device: such a desire has to do with life style and identity. This is much like other ICT research. There is, however, a little scientific data on the emotional character of the use of mobiles across Europe: particularly for the emotionally preoccupied (teenagers), but it does not say much about what that emotional character might be.

Nonetheless, combining all the evidence that there is makes it clear that mobiles are experienced in a way that is distinct from other ICT.

7.3.3 IMPLICATIONS FOR UMTS SUPPLIERS

Using emotion in the way described here to maximise the potential of products and services is challenging not least because it demands a level of understanding of customers – existing and potential – that is not normally known. More particularly, it demands knowledge of the purpose of the user's communications and how those purposes deliver the emotional value that is so important to them.

The use of SIM card readers by some Service Providers to collect and store information held on mobiles – phone numbers in particular – is a present day example of a service that responds to users' fear of losing data but goes only a small way to addressing how to do so in a way that reflects the emotional value that same users place on information which they may lose. In other words, much more is needed to be understood about how emotional values are delivered and preserved before UMTS providers can identify ways of leveraging opportunities related to emotion.

Failure to understand these emotional values can lead to problems. New services could be jeopardised if they replace or impinge upon services to which the user ascribes an emotional attachment. For example, the threat of losing text messaging and having it replaced by new technologically better services may create considerable resistance. For not only is texting now a key tool in sustaining emotional lives, but storing personal text messages is now a highly valuable element to people's emotional arsenal. SMS may be thought of as simply a communicative technique from the supplier's point of view, but to the user it has values over and above this.

Relatedly, the adoption of new form factors may affect these emotional values. For example, the current size of GSM devices supports constant carrying around: 'it never leaves you', as one of our subjects put it. This means that users are never forced to relinquish contact with those they need to be in contact with. Now, it should be clear that before GSM they would have had to be out of reach for certain times of the day and in certain places. The point is that with GSM they have come to expect this constant access.

Future devices must not threaten this. Many UMTS services and products require larger screens than most current GSM devices require for example, and this may lead to expanding the form factor to a level that makes constant carrying difficult or at least irritating and burdensome.

Similarly, one-handed input means that it can be used at almost any time 'even when driving'. Many smart phoned devices, particularly those which combine the PDA form factor with soft keyboards for dialling and so forth, require two handed input. This inhibits the places in which they can be used. Clearly in some respects this could be an advantage: stopping people using the mobile while driving may be viewed by some as a way of increasing safety on the road, for instance. But for many users it is precisely the ability to be in touch at any time that provides the value that has made the GSM phone distinct. Assault these values with new UMTS products, even if it is only at the edge, and the overall value may be diminished.

Beyond these specificities of form factor and service change inertia, a further and perhaps more significant lesson is that demand for services that sustain emotional lives may be highly inelastic: people may pay 'whatever it costs' to have and use a mobile phone (though the cheapest will do). The reverse side of this is the Value Paradox, by which you mean that the value of the mobile makes losing it too risky. More importantly, other technologies (e.g. the home PC) do not gain the same value. In this sense, the threat of convergence of the PC, the PDA and the mobile may not be real.

The findings also have implications for the unpacking of WWW data from 'words' (voice and text). Person to person connectivity services engender emotion; this is where the value lies. In contrast, 'person to WWW data' does not achieve the same emotional value. This resonates with the previous findings on social connectivity, and indeed gives further weight to the importance of person to person connections as against person to information. According to both the analysis of the emotional relationship people have with their mobile and the person they connect to, the implication is that money can continue to be made through developing person-to-person connections, and much less from person to information connectivity.

7.3.4 INTERVIEWS WITH EXPERTS: REVISIONS OF THE IMPLICATIONS

The response to the thesis was generally confirming, though the extent to which the relationship with mobiles could be described as emotional varied according to cultural differences.

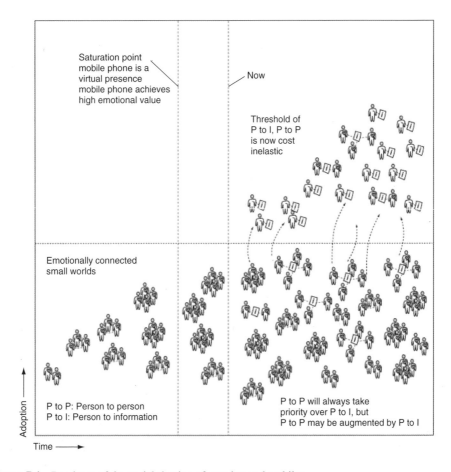

Figure 7.4   Roadmap of the social shaping of emotion and mobiles.

Explanation

The roadmap shows that the consumer marketplace is made up of masses of emotionally connected and interconnected small worlds. At the current time (the bottom left hand side of the figure) there is little person to information use. As UMTS products appear, however, they will need to be based on these emotional ties, and achieve some augmentation of person to person connectivity with some person to information connectivity. The figure also shows that providing person to information connectivity may be difficult since there may be greater price sensitivities for such services than for person to person connectivity.

Some readily agreed that it was clearly emotional, while others said that they agree the relationship could be thought of as emotional even though users were unlikely to coin that phrase themselves.

All agreed, however, that such things as the personalisation of GSM devices (colour, directories, ring tones) was less a reflection of the normal pattern of the evolution of consumer products (where personalisation to some degree is used to differentiate products) as it was an indication of the particularly emotional character of the user's attitude toward mobile devices. Some even remarked on the way mobile devices are held and touched, remarking on how mobiles are sometimes charged with almost sensual properties to affirm their belief that the relationship with mobiles is indeed different, cultural differences of expression notwithstanding.

Looking toward the future, the expert panel commented that, in the first instance, they will attempt to leverage emotional values to identify and sell new products. Key to this will be using emotional relationships between friends as a route to offer services and products that augment those relationships. Gradually, and in the longer term, the panel explained that they would introduce products and services that would be less and less emotional, thus leading themselves out of the confines of satisfying purely emotional needs. Their view was that they need to start developing UMTS services through leveraging person to person emotional needs, and then, step by step, introduce more person to information like services (Figure 7.4).

7.4 THEME 3: PUBLIC AND PRIVATE

According to many scientific commentators and recent research conducted by the DWRC, GSM mobiles created and continue to create a blurring of social codes of mobile usage for public and private spaces. After the initial shock (in the early nineties perhaps), now almost 'anything goes': people will use mobiles for any type of communication anywhere. However, the evidence also suggested that new products and services might be a 'step too far'. In so doing, new products may lead to social and regulatory resistance, with signs saying 'No mobiles here please!' becoming more common.

To investigate this, the Study combined state of art research plus new first hand evidence to explore family, social and business use of Location Based Services (LBS) and perceptions of imaging on mobile phones. New evidence was gathered through focus groups in Germany and the UK. These data were then compared and contrasted with data gathered through direct observation, given that prior research has repeatedly demonstrated that what people say (in focus groups especially) is not always what they do. Findings from the research indicate that the future for each type of services is potentially quite different, with different aspects of technology design and service provision impacting upon take up. More particularly, LBS and imaging were treated as distinct topics of inquiry.

7.4.1 LOCATION BASED SERVICES

Location based services are already in use and new services are being introduced at an increasing rate. The Study asked people what they thought about location based services in general rather than about specific products and it mostly considered tracking, locating people and advertising as the service principles.

7.4.1.1 Findings

In focus groups, both in the UK and Germany, the Study found that the dimensions of tension listed in Table 7.2 are used to account for consumer views on the topic of LBS; the top two are the most salient.

Focusing on the last dimension, family life, it became clear that parents felt that tracking teenagers would be 'terrible', 'offensive', 'horrible'. This was a recognition that the benefit of tracing children in trouble would be countered by an awareness that overprotective parents could misuse the facility. Imposing artificial location limits on teenagers were also deemed unacceptable, since 'it takes away your (e.g. children's) ability to cope'.

Paradoxically, the perspective of teenagers and children was quite different. Though they admit that they would turn off or throw away their mobile phone to avoid tracking by their parents, teenagers like being able to locate their friends, and most teasingly of all to parents, teenagers like to be able to locate their parents.

Moving away from the family and thinking about LBS more generally, trust and distrust had to do with things like how the information would be used: one person summed up the ambience that LBS might create: 'you'd get paranoid', and interestingly, this would apply to both the user and the subject. If a third party were used to mediate location information (i.e. a third party accessing information for the operators) there was a concern as to who this should be: the view of subjects here was not that it might be the local curry house – an example often cited in the LBS literature – but rather Big Brother.

In this light there was some resignation that LBS would be like CCTV, with the view that 'This may happen anyway and people will just get used to it' being heard several times. A key difference, of course, being that CCTV is in spaces already defined as public, whereas with LBS, what is public and private is blurred.

Concerns raised with the business use of LBS had, first of all, to do with similar issues: namely, coping, reliance and utility. 'It's telling me stuff I don't need'; 'Sometimes I do need information, but how does it know?'; 'What would I do with it anyway?' reflect the views

Table 7.2 Summary of focus groups' concerns about Location Based Services

LBS	Explanation
Coping vs. reliance	which has to do with managing oneself versus relying on others
Trust and distrust	and the question of ethics, conflicts with freedom and learning social skills
'Big brother' and motive	which labels a concern with who is using LBS information and how trustable are they?
Hiddenness vs. explicitness	which has to do with who one might be willing to tell about one's location
Exceptionality	which reflects a concern that LBS would only be useful in certain cases
Push vs. on demand paradox	where there is a recognition that LBS may provide information that is not needed or not providing access to information that is
Family coping and reliance	which has to do with complex relationships of dependence and independence characteristic of families

expressed. Combined with this was concern that as with parents and children, LBS might have the effect of reducing people's ability to cope, despite the fact that 'They have got to get on with it by themselves'.

Besides, subjects made it clear that people do not like the idea of being checked on themselves nor automatically knowing where employees are. Both were viewed as intrusive: 'people don't give their best if they think you're on top of them'. Tracking or limiting staff within an area is also unacceptable, 'It would be like saying you don't trust any of your staff'. The best overview of this concern was expressed in the following way: 'Because Technology CAN do something it doesn't mean it should'.

7.4.1.2 Current patterns of use

Though the focus groups' data was rich with insights and issues, the DWRC team juxtaposed that data with what is known about mobile behaviour. Behavioural findings show that the use of mobiles is, as it were, 'saturated' with location sensitivities. Yet it also shows that though mobiles are used to connect to 'any place, any time', this benefit or affordance is managed.

For example, people start mobile phone calls with 'Where are you?' That they do so demonstrates that people negotiate the right to intrude. Another example of such negotiation, sometimes tacit, sometimes explicit, can be found in the use of Text after 11pm so as to allow the recipient to allow or not allow interruption, or similarly the use of text when the recipient is in a meeting.

In other words, the reaction to LBS scenarios in focus groups and other research needs to be placed in a larger context, in particular, in regard to the fact that users currently manage levels of intrusion enabled by mobile communications in subtle but nonetheless conscious and efficacious ways. This demonstrates that location does matter and key to this management of this concern is recipient control.

This is itself complex and is attested to by the fact that even those who know the recipient very well (and who may thus have a good idea of where that recipient might be), typically let the recipient make the decision as to what is an appropriate level of intrusion. This is allowed through the negotiations that commence a call, as already mentioned, or through the use of discrete connectivity media such as SMS again already mentioned.

An additional factor that has to be taken into account has to do with the fact that it is largely friends and family who are using their mobile phones to contact one another, and that even in business, mobiles are used for communicating with colleagues rather than with clients (a proposition examined earlier in this report in relation to social connectivity).

Given these two facts, then, there are two questions for LBS. Firstly, whom will be the party providing the service? If it is a business, rather than a consumer pull process, then a service provided by that party might meet with resistance. This is because the services may be more desirable to the business providing the services than to the customer they are hoping to reach.

Secondly, what relationship do the providers of LBS have to friends, family and colleagues? Are the users in control or the operators? The implication is that if it is the business rather than the consumer, then this is not likely to be acceptable. In a phrase, it may be that business to consumer (B2C) LBS will not succeed.

This is not to say that LBS is unlikely to develop; what it suggests is that for this to happen features of current practice need to be leveraged. This can occur once it is recognised, on the basis of the above findings, that there is a balance between need and social etiquette.

Moreover, friends and family do have social structures that might support LBS: Parents want to locate children but children resist; teenagers want to locate other teenagers and parents but not let parents locate them. The issue, then, is not the need as it is already there: the issue is how to address the need.

7.4.1.3 Implications for UMTS Suppliers of LBS

The key dimensions of relevance for LBS, then, reflect subtleties in mobile communications behaviours, subtleties that can easily be overlooked (and indeed are by most of the suppliers in the market place). There is the fact that consumers find the idea of LBS worrying. Second, there is the fact that mobile usage is, nonetheless, saturated with location issues (from the users' perspective): consumers orient to intrusion of public and/or private space through a usage 'etiquette' (e.g. texting at certain times and places). And finally there is the question of who is being connected with the mobile: currently it is close friends and family but not business to consumer.

The Study takes the view that the UMTS suppliers need to avoid presenting LBS to the consumer as somehow an abstract concept. If they do present it in this way they will meet with the consumer rejection as described. Instead they need to design 'control for the recipient' into LBS. In particular they need to look at the 'opening sequence' of communications between users, showing an especial concern with how various types of messaging services are commenced with distinct but nonetheless rule bound opening sequences (by rules the rules of social etiquette is understood). The most important of these has to do with what is called (in the sociological literature) the 'summons–answer sequence', where a receiver of a call (i.e. the one summoned) has to respond in an appropriate manner. What is found with mobile communications and messaging in particular is how the summons–answer sequence is altered insofar as the recipient has more control over whether to respond and what to say in response than the summoner, i.e., the one making the call.

According to this view, LBS will have greater success if it allows users to opt for products and services on an expressly 'need to have now' basis. This might be according to time and place, and would be on an entirely ad hoc basis. Thus permission based products that require users to accept LBS as part of a complete unalterable package (what might be called a 'carte blanche' approach), that is intended to secure customers through offering such things as tariff reductions for rental are, according to this view, unlikely to succeed, whatever the inducements. They would be perceived as a 'step too far'.

Key to delivering such services is designing an appropriate MMI that provides the recipient control as mentioned. Despite the evidence from the scientific research that the usability and appropriateness of the MMI is likely to be key, little consideration of usability factors would appear to have been given by suppliers of LBS services or indeed the operators who would provide data for those services. This it would seem is potentially dangerous neglect.

Irrespective of whether sufficient attention is given to the MMI, UMTS suppliers of LBS also need to design for social connectivity: for friends, family and colleagues. The evidence shows that services that succeed will support consumer to consumer and consumer to business and less so business to consumer services.

If suppliers fail to take the issue of who will use the services and the interface to the devices seriously – perhaps trying to design around them by making their services in such a fashion that matters of social etiquette are treated as irrelevant, for instance – then they will find, according to this analysis, that 'doing' LBS without looking at social factors will kill the opportunities that would appear to be there.

7.4.1.4 Interviews with Experts: Revisions of the Implications for LBS

In general, the interviews with the expert panel confirmed these arguments. In particular there is a common recognition of the strong consumer suspicion about LBS and a need, therefore, to be very sensitive about their introduction.

Nonetheless, there were some dimensions along which there was much less consensus. For example, though there was a general conviction that resistance can be overcome with the right incentive, particularly for permission-based LBS, there was no agreement to what this incentive might be.

In any case, the expert panel noted that there are so many types of LBS that counter examples could always be provided for any attempt to identify social shaping drivers beyond basic consumer suspicion. Indeed, the multiplicity of products makes production of a social shaping roadmap for LBS impractical.

7.4.2 IMAGING AND MOBILE PHONES

As with LBS, the concern with this topic was to inquire into whether imaging is a step too far. The data gathering and analytical process was the same as with LBS.

In summary, the Study found that consumers find the concept mysterious: they are not sure how to use it for communicating, and they are sensitive to loss of privacy. They are, however, excited by possibilities. Moreover, current behaviours show a willingness to adopt the new medium and new usage patterns, though for success in the longer term the analysis suggests that this new usage pattern should be similar to that related to text. These patterns of usage are more important than the fact that consumers like to use the mobile phone as a watch, radio, reminder etc. and therefore the imaging function might be a nice add-on feature.

7.4.2.1 Findings

The focus groups identified a number of dimensions key to understanding the future social shaping of imaging. They are shown in Table 7.3.

Focusing on each in turn the Study found, with regard to togetherness and novelty, that the opportunities for using mobile imaging are quite exciting: 'You could take a picture and send it to your mate'; or it could be used to say 'Look what you are missing out on', 'My sister would love it – she's nine'. Nonetheless, there was a concern that 'You'd get bored of taking pictures'.

A paradox was perceived, however, and that had to do with appropriateness of the technology. The model users have is of an image quality that is as good as the more traditional photograph, i.e. one where the imaging quality would be very high. Thus attitudes to usage and costs associated with it reflect what was assumed to be the purpose of current photography

Table 7.3 Summary of concerns about imaging over mobile phones

Togetherness	Family and friends sharing moments
Novelty vs. reality	Fun at the beginning and then get bored
Secrecy	Moral issues of taking covert pictures, possibility for abuse
Complexity and quality	How difficult and how good will it be
Appropriateness	When it's not and when it is OK

and, given this, that there might be little fit with current usage motivation: 'Mobile phones are for day to day low level operation, photographs are for special occasions'.

A further concern had to do with the fact that cameras could be used to photograph people without their knowledge: a response often expressed to this concern was the suggestion that the mobile imaging device 'must look like a camera'. Potentially sinister usages were not the only worry: there was also a concern that 'you could send an unsuitable image to your Mum by mistake' (experience learned from text!).

There was some expectation that incompatibility of mobiles and networks may cause difficulties sending photos to friends 'We've all bought Nokias so we can exchange photos – it doesn't work if one of us has a Siemens'. Many expressed the belief that a download facility to PC is needed to make it worthwhile. This reflected a concern about inadequate size of screen and a problem of where to store the images.

7.4.2.2 Current Patterns of Use

Unlike LBS there is no social etiquette for taking and sending instant pictures with mobile imaging (though there clearly is already one for taking photos). This leads to a question as to how imaging products could be used: subjects asked themselves when they would use it and were unsure. They were also concerned with what they viewed as the practicalities, which they viewed as insurmountable. Partly this had to do with the model of a similar technology they had in mind, where cost was known to be high and technology challenging. (What they were thinking of here is both SLR and digital photography.) If the model being used mirrors the one used with these more traditional technologies, then subjects raised concerns about quality and size of the image with mobiles, which they viewed as being poor on both counts. Even so the idea was indeed appealing: for recording special family moments or proof of an event: 'I saw Jude Law in Covent Garden but no-one believed me'.

7.4.2.3 Implications for UMTS Suppliers of Imaging

Imaging over mobile phones is in its infancy but imaging (camera, cinema, video etc.) has an historic legacy in peoples' day to day lives that is influencing their willingness to adopt imaging as an augmentation to a mobile phone. However, the novelty and complexity of imaging over mobile phones at the current time clouds various issues to do with what this process of augmentation will lead to.

For example, people are concerned about such things as appropriateness: 'Is this really what a mobile phone is for?'

They are also worried about secrecy: 'Who has taken a picture of me (my child, my innovative design) and what are they doing with the image?'

Products that assuage these fears will succeed only if they can shape consumer expectations in a certain way. Given that phone imaging is still in its infancy, the suppliers still have the opportunity to make sure that this is the direction that social shaping takes.

More particularly, according to this analysis, suppliers need to show how mobile phone images are for friendly, 'in the moment' and largely ephemeral communicating and not for replacing 35 mm or digital photos which provide lasting memories. In a phrase, they are not for ceremony but for pictures to play with.

This means that suppliers need to put effort into managing down expectations for quality and storage of images among consumers. A model could be text, and so the products should be encouraged to be used like that medium, which means as additional to such things as the traditional post card and digital camera image of special events. According to this vision, the goal would be to support visual 'chitchat' not wedding photography type of memorabilia.

If this is the case for imaging, the issues as regards the social shaping of video telephony are much more substantial. Here the need to embed the problem of managing intrusion with fully two way video are quite large for users. Only through embedding extensive recipient control in to the devices themselves will video telephony find acceptance.

For example, video-telephony will not be acceptable except for a small proportion of users if there is no capacity for the recipient to negotiate the choice of video, talk or text at commencement of each communication. This has to do with what was remarked upon earlier regarding the need to ensure recipient control.

This control has a number of salient aspects: one is that a video caller should not be able to switch on the camera at the recipient's end; rather it is only the recipient that should be able to do this. But currently the way the system is designed to work a video call is either 'go or no go' with capacity to start with talk and then move up to video and people might want to change after an initial hesitation (though some operators are beginning to alter this).

Similarly, a camera should routinely be facing away from the recipient rather than towards, so that they can more effectively control what is being seen. There is a belief that recipients are more likely to show what they can see (i.e. what they are looking at) than let the caller see their faces. This has in part to do with concern about what a camera might see when placed close to the ear and so on. Hands free solutions to this dilemma are likely to be dismissed since they seem to create further problems of managing the public and the private with many subjects expressing a fear that with hands free it becomes more difficult to ensure privacy.

7.4.2.4 Interviews with Experts: Revisions of the Implications for Imaging

Unlike the case with LBS, there was almost complete consensus about the implications for the evolution of UMTS imaging services with the expert panel.

First, there was a consensus as regards imaging as being an application suited for 'play' rather than for serious or as it were practical functions. It will thus flourish in the way that text has. Managing consumers' expectation about the quality and role of imaging will need to be handled carefully, however, if this is to be ensured.

As regards video telephony, there was consensus that there will need to be a substantial shift in levels of usability for imaging technologies to become more than niche products (Figure 7.5).

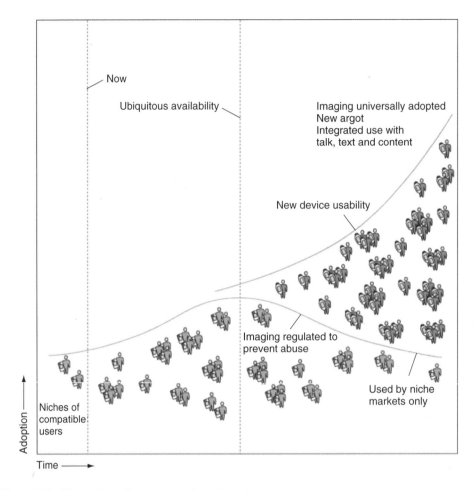

Figure 7.5   The roadmap for social shaping of imaging.

Explanation

Currently, imaging is only possible between groups of niche users. Without substantial change in MMI, particularly for video telephony, however, there is a strong possibility recognised throughout the industry and the scientific world that regulatory resistance may show itself: and if it does not then certainly the technology may end up being confined to a small group of users. With changes in the MMI substantially more users will adopt the technology.

7.5 HOW GSM WAS SHAPED BY THE CONSUMER

7.5.1 INTRODUCTION

Although much has been written about the GSM, most of it does not address the subject in a way that would be useful for understanding how people might respond to UMTS. Instead, it treats the subject largely in technological terms. In addition, the speed of take up, the development of products and the growth in customisation was (and indeed remains) rapid, and this has resulted in the assumptions underscoring the basic GSM mobile communications technologies and products becoming increasingly varied.

Nonetheless, there is an increasingly large literature in the area of 'user studies' relating to the use of mobile phones (GSM and otherwise), and during Phase one of the research the Study gathered together research insights from this literature and combined them with the DWRC's extensive data sets on user behaviours and social shaping. The Study then selected those hypotheses from the literature and research that appeared potentially significant, and developed various empirical activities that validated and deepened some of these research insights in Phase two.

Methodologically, the approach used throughout was to focus on mobile communications from the perspective of the GSM mobile phone users. It was thus qualitative and involved reviewing anthropological, sociological and psychological studies of human behaviour, as well as 'ethnographic' research conducted as part of the Study. In benchmarking the present from this perspective, the Study was able to learn how GSM devices have shaped and have in turn been shaped by fundamental observable patterns of behaviour in society.

It should be clear, however, that the use of qualitative data produces insights and analysis that may not be familiar to those organisations planning to deliver UMTS services and products. They may be more familiar with the kinds of techniques – primarily quantitative – used in engineering modelling and marketing based studies. Nonetheless, the kinds of insights that the research could deliver are sufficiently important that providers of UMTS services and products must attend to them irrespective of their familiarity with the data gathering and analysis techniques.

7.5.2 EVIDENCE OF SOCIAL SHAPING – BENCHMARKING THE PRESENT

More particularly, the literature review undertaken in Phase one commenced with a judgement as to scope: though as has been said there is little literature citing GSM and social factors, there is in fact a great deal of literature on social shaping. In order to set some limits, the Study opted to demarcate a historical point before which it would not look.

This was the point at which mobile telephony became omnipresent. In the UK and the rest of Europe this would have been the situation in the late 1980s/early 1990s where the analogue TACS/NMT mobile phone services had been available since the early 1980s and the new GSM digital service was launched in 1993. Elsewhere, analogue AMPS service was prevalent, being upgraded to digital and then replaced by GSM by the late 1990s. The global adoption of GSM meant that for many nations GSM was their first cellular mobile phone service and coverage was usually limited to major conurbations, albeit quickly extending nationally and often providing coverage where fixed line service was not available.

What the Study found in its research was that although people quickly became familiar with the concept of digital mobile telephony, it was at the outset high priced and targeted at business users although some individuals did adopt it for personal use. The regulatory frame-work of the time was created to address the issue of price, of course, and leveraged competition. In a crude sense this was social shaping of the most obvious kind.

The UK market, for example, was structured as a 3-tier system and this meant that people had to buy their mobile phones from a Dealer/Retailer who was selling on behalf of a Service

Provider who bought wholesale from the Network Operator. This prompted a new way of selling and buying telecommunications and a new economic structure of incentives and subsidies.

Initially the mobile phone did not actually displace other communications devices – as new systems were introduced they replaced and updated antiquated mobile telephony services but, overall, it was an addition to people's lives. People had not previously been able to make voice calls on the move, in locations where there was no fixed phone or have a number personal to them. With GSM they could do all of this.

The technology enabling mobile telephony thus certainly shaped people's lives insofar as it made these possible, but once they had leveraged these capabilities what changes followed on? Did the social shaping continue?

The answer is a resounding 'yes' but the blurring of who (or what) was shaping whom increased as the mobile phone became more embedded in peoples' everyday lives, and technological developments continued to create new capabilities.

How then can the research unravel this web of interdependencies? What the Study found is that there were some common themes relating to social shaping and mobile technologies, themes having to do with change at an individual, organisational and social level. In brief, they are:

- *Patterns of adoption.* There was a clear pattern of adoption of GSM that is common with the introduction of other new technologies such as broadcasting technologies (e.g. TV). Though most agree that one should question whether this would re-apply to UMTS, there is much less agreement as to what the alternative path of development for UMTS will be.
- *Ecology of mobile phones.* Mobile phones created a new ecology of technologies and communications, and have thus became a key aspect of work, family and personal life.
- *Different type of culture and social groups and their relationship with providers.* There was clearly a complex relationship between the culture of a customer and their provider, mediated through not only the devices themselves but also such mechanisms as billing and payments methods. Thus for example, prepaid was a success in Europe and less so elsewhere; there were cultural differences in how consumers viewed content relative to service and terminal providers; and so on.
- *Developing form factors.* There has been more than simply shrinkage in the size of mobile devices and the evidence suggests that the success of some terminal manufacturers would appear to be an outcome of a combination of technological development and consumer needs matching one another.

Within each of these themes there is a cluster of sub-themes that encapsulate the social shaping drivers within society that affected the evolution of the GSM. In particular, and beneath the above listed themes, seven main issues or hypotheses are identified.

1. *Social relations.* To begin with, there seems to be a consensus in the scientific community that mobile phones had – and continue – to reinvigorate social relations through providing a voice or text mediated form of face-to-face relations. Some commentators view this 'virtual presence' as counterbalancing the increasing social isolation created by other new digital media, such as interactive digital TV, computer gaming and the Internet. This benefit made GSM communications unlike other digital technologies and unique from the users' perspective;

2. *Personalising*. It was also argued that with GSM, person to business relationships became much more personalised than before, with mobile communications allowing more intimate and frequent contacts;

3. *Particularisation*. Another thesis was that mobile networks provide much more fine grained, 'particularised' information about user behaviours than was been possible hitherto, though users did not – and still do not – perceive this as a concern nor did business effectively leverage any opportunities this provided. Much of this data has remained untapped (though new services are likely to latch onto its possibilities – location services, spam text and so on);

4. *Emotion*. Many commentators suggested that the relationship between the user of the mobile phone and the device itself became much more emotional with GSM devices than was hitherto the case with digital technologies. It was argued that this was a function of the social connectivity that mobile phones afford and reflected a relationship with the content delivered via the device more so than the device itself;

5. *Private behaviours*. Another view was that mobile phones resulted in more private behaviours in public spaces than ever before, with gradually fewer boundaries to acceptance of where and when people could use their mobile phones. This was a worldwide phenomenon, though the extent to which it occurs varied between different countries;

6. *Mobile interaction*. Though the evidence was sparse, it was also argued that there was an increasingly distinct form of interaction on mobile devices, which involved frequent, short duration, and low content needs. This emerged partly because of network quality and cost and partly because of new social behaviours where chit chat by voice and text began to replace more focused, lengthy communications behaviours;

7. *Pattern of augmentation*. Finally, the research on GSM demonstrated that mobile phone users were in effect 'in charge' of the process of augmenting their mobile applications and services, and this was reflected in the way they chose to adopt some services and products and not others; disregarding the attempts of the suppliers to do otherwise.

7.5.3 FURTHER EVIDENCE OF SOCIAL SHAPING

Building on the main areas of research on the social shaping of GSM and these deriving hypotheses, three hypotheses in particular were then chosen following consultation with the UMTS Forum. Phase 2 concentrated on the specific characteristics and implications of these hypotheses about social factor drivers and the implications they might have for UMTS products and services. The relationship between the selection of these and the general literature on social shaping is represented in Table 7.3.

Clearly, social shaping will have many more facets than could be covered by three themes though this number were selected and these in particular because they were viewed, by the Study and the UMTS Forum, as sufficiently important for the development of 3G to warrant closer examination and new data gathering.

The hypotheses were:

1. *The intersection of public by private behaviours enabled by mobile phones will reach a threshold beyond which resistance will start to occur*. Prior research had suggested that public spaces are being invaded by private behaviour but that at a certain point resistance

to this change would occur. Two specific areas of private and public behaviours were therefore explored, in relation to this issue. The first was the potential use of location-based services; the second was the use of mobile images.

2. *Mobile devices do not enable more social relations but more intensive relations with already existing contacts.* Prior research has suggested that mobile communications are

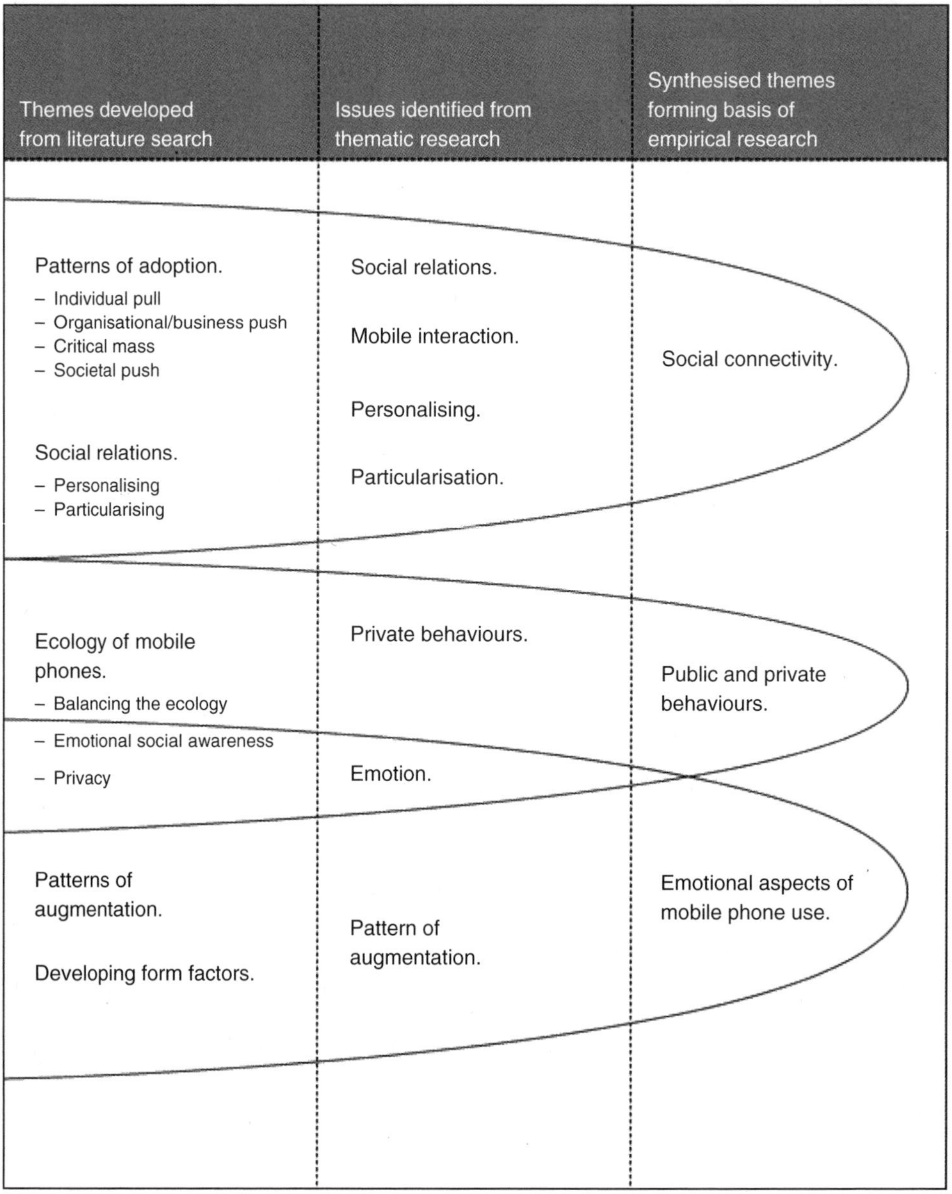

Themes developed from literature search	Issues identified from thematic research	Synthesised themes forming basis of empirical research
Patterns of adoption. – Individual pull – Organisational/business push – Critical mass – Societal push Social relations. – Personalising – Particularising	Social relations. Mobile interaction. Personalising. Particularisation.	Social connectivity.
Ecology of mobile phones. – Balancing the ecology – Emotional social awareness – Privacy	Private behaviours. Emotion.	Public and private behaviours.
Patterns of augmentation. Developing form factors.	Pattern of augmentation.	Emotional aspects of mobile phone use.

Figure 7.6   Flow chart showing the capture of the data and synthesis into the 3 core theses.

creating changes in social relations. The first element of this thesis is that mobile communications are increasing the degree of connectivity between people, but that this connectivity does not increase the numbers of people that are in contact with each other, rather it increases the frequency of contacts between the same people. The second element of the thesis is that mobile phones are creating an opportunity for new and informal relationships in business in particular.

3. *Users have a more emotional relationship with their mobile phone than they do with other forms of computational device*. This thesis did not indicate that users did not have emotional facets to their relationship with other technologies; it was rather that on a scale the character of the relationship with the mobile was more emotional than any other. Moreover, this resulted in certain kinds of behaviour that one does not find with other technology.

The taxonomy of the links from initial research with the 3 themes that form the basis of the study are represented in Figure 7.6.

Accordingly, in addition to examination of the extant literature on the topic, some new empirical research was carried out in relation to these themes. These data gathering exercises comprised 10 focus groups and a questionnaire. This research was conducted simultaneously in Germany (Erfurt) and in the UK (London and South East) using the same brief for the focus groups in both countries. This was a largely qualitative study, seeking people's feelings about their mobile phones, their use of them and what they thought of some of the new services that might become available in the future.

The Study then analysed the data, combined with other newly reported scientific research, to develop its interpretation of what the implications of these social shaping factors would be for UMTS suppliers. The Study then presented its analyses to key players in the mobile industry and refined and revised its views accordingly.

Each of these themes and the resulting implications, including the revisions generated by discussions with experts is presented in the body of the report.

8

3G and Beyond[1]

8.1 A LONG-TERM BUSINESS ISSUE

Much has been written about revolutionary new capabilities and evolutionary development. A few, such as the WCDMA air interface of UMTS and the IP Multimedia Subsystem, represent revolutionary advances in capability. However, the approach is always to provide for the enhancements by adding platforms to existing networks and by upgrading existing platforms rather than discarding equipment and starting over again. This applies particularly to the GSM core network, which remains the basis of the UMTS/3G system throughout the evolution and will play a dominant role in 3G and beyond (Figure 8.1).

Evolution of UMTS is already being considered. The 3G/UMTS radio access technology will be enhanced to support high speed downlink and uplink packet access, enabling transmission at speeds of up to 10 Mbps. There will be other complementary technologies in order to provide really high data rates and very high user densities, such as would be found in conference centres, including wireless local area networks (WLAN) as was described in Chapter 4, which will complement UMTS in future, offering theoretical bit rates up to 55 Mbps. Although public WLAN networks will also be deployed independently from the mobile networks, there are built-in advantages for the mobile operators that come from the ability to provide mobility management, subscriber management, high security and roaming.

Figure 8.2 describes a possible evolution of access technologies for 'systems beyond 3G' to reach 4G. A technology, orthogonal frequency division multiplex (OFDM), already discussed for 3G with limited success, gets a new chance. It has some advantages for very high speed transmission (~100 Mbps) and seems to be a very promising access technology candidate for 4G. Laboratories, not only in Japan but also in Europe, e.g. Institut für

---

[1] Professor Barry Evans generously permitted the author to reproduced the figures he used at the DWRC 5th World Wireless Conference for this chapter.

The Mobile Multimedia Business: Requirements and Solutions   Bernd Eylert
© 2005 John Wiley & Sons, Ltd

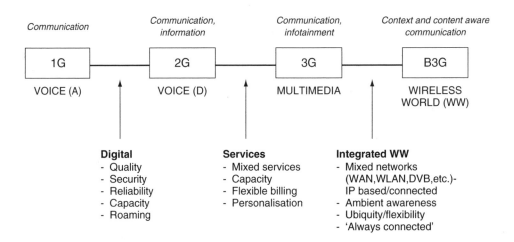

Figure 8.1   The mobile communication evolution to 3G and beyond. *Source*: Evans (2004).

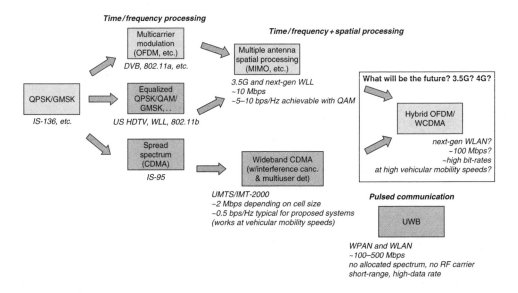

Figure 8.2   Modem technology trends. *Source*: Evans (2004).

Nachrichtentechnik of the Technical University Hamburg-Harburg and the Bosch Laboratories in Hildesheim, Germany, have been very active in this research for many years and it seems to be bearing fruit today.

   Another development path is given by ultra wide band (UWB), an evolution path from WLAN and a member of IEEE's 802.xx family. As a big milestone on this way the product WiMax, IEEE standard 802.16, was ratified in June 2004 and is expected for certification in 2005.

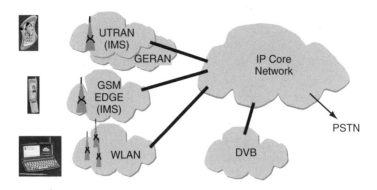

Figure 8.3 The interworking networks. *Source*: UMTS Forum (2002).

The technology world remains very interesting and it seems to be quite likely that a new 'technology war' may show up, as happened with the evolution of 3G.

Another enhancement is the IP multimedia subsystem (IMS), as described in Chapter 4. It enables real-time, person-to-person services, such as voice or video telephony, to be provided by means of packet-switched technology in common with information and data services, by using IP multimedia call control. It allows the integration and interaction of communications and information services, as well as enabling communications sessions to be established simultaneously between multiple users and devices. The introduction of IMS parallels the revolution that is taking place in fixed networks in the form of voice over IP (VoIP).

Yet further advantages will come in the longer term from the ability to interwork inter-actively with other networks such as digital video broadcasting (DVB) and take advantage of the content offerings that can be delivered efficiently to small form factor mobile devices (Figure 8.3).

8.2 THE RESEARCH AND DEVELOPMENT PATH

There is a stream of technologies emerging from research and development (R&D) with different time horizons (Mobile Communication and Technology Platform, 2004). Some find use immediately in present systems and applications, while others can only be implemented after other technological breakthroughs or changes in demand, regulations or standardisation occur. Standardisation and regulation tend to be set at a given time ahead of market introduction and then remain relatively stable before they are adjusted to an accumulated pressure for change as products mature and require new functionality.

This leads to several concurrent streams of activities, which interact with each other. While not always apparent, regulations may shift the priority for R&D and the setting of standards may make promising technology options irrelevant for some time to come. Conversely, standards can become irrelevant when R&D opens up new options, resulting in de facto standards driven by demand.

As this will be evident from the many examples of the work of the UMTS Forum which are referred to in this book, industry has been working together very closely for some years. Research and development of these systems took place in the decade between 1991 (the time when the first GSM systems were launched) and 2003 (the time when UMTS was launched in Europe). Research on enhancements to third-generation systems is already very advanced in 2004 and is entering the stage of handover from research to product development.

Industry is committed to continuing R&D cooperation in the pre-competitive domain. This is especially the case for 3G and beyond. Therefore, the Wireless World Research Forum (WWRF) was launched in 2001 on a global basis. The objective of WWRF is to formulate visions on strategic future research directions in the mobile and wireless field, among industry and academia, and to generate, identify and promote research areas and technical trends for mobile and wireless system technologies. It is intended to construct-ively contribute to the work done within other bodies regarding commercial and standard-isation issues derived from the research work. The forum is open to all interested parties. The book *Technologies for the Wireless Future* by the WWRF provides an overview of the research issues that need to be investigated to develop the wireless world (Tafazolli, 2004). Figure 8.4 is an illustration of the steps WWRF will undertake to synchronise its work, split in to milestones, with the corresponding work programme of the ITU-R on 'Systems beyond 3G'.

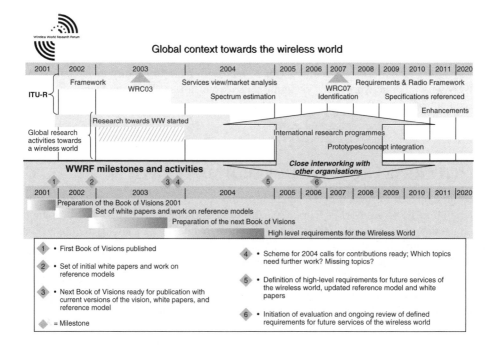

Figure 8.4   Timeline of the Wireless World Research Forum. © WWRF.

8.3 RESEARCH PROGRAMMES

8.3.1 ASIA

Compared to Europe and North America, the countries in Asia, especially China, Korea and Japan, started very early on new R&D programmes on 3G and beyond, quite often called 4G. They established '4G Future Groups', started a search programme on finding new spectrum for 4G in the range of 3–6 GHz, and started research programmes on broadcasting HAPS/ satellite and the future 4G standard for 20–100 Mb/s. These countries for industrial-political reasons are very keen to speed up with 4G to set the scene on this evolution. Historically, with their 2G development the leading Asian industrial countries were not commercially successful on the world market. However, Japan in particular could play a dominant role in 3G and was eventually successful with 3G, because it decided to work very closely together with Europe. The starting role was taken by NTT DoCoMo when it started intensive cooperation with leading European companies. Having reached this position, it will continue with the same strategy on 4G to keep its dominant position on R&D in telecommunications. There is no other country in the world that has put as much effort, money and scientific power into the project 'Systems beyond 3G' and 4G as Japan. Korea closely follows its Asian rival Japan, and China, with the world's biggest market, does its best to keep up with them.

8.3.2 USA

With growing interest of the US industry and administration in 3G, the industry bodies in the USA started some research on 3G and beyond, but mainly focusing on activities in the unlicensed bands 2.4 GHz and 5 GHz to support the ongoing work on WLAN/hotspot and push for the IEEE 802.xx standard. Some market players already identify WLAN as a 4G technology, but they also recognise that this is not enough to fully define a new mobile generation.

8.3.3 EUROPE

In Europe, a range of research projects on concepts and prototypes for systems beyond 3G is already under development, both in the context of private investment and in the context of the 6th Framework Programme of the European Commission. This research will lead to exploitation and initial products in the market in the period 2007–2012. A review of the outcome of the 5th Framework Programme can be found on the IST Programme web site (see www.cordis.lu/ist/cpt/ippa.htm). This must also be seen in the context of the proceeding works WWRF has already undertaken to synchronise Europe's voice in the world's mobile community. Figure 8.5 illustrates the R&D work programmes undertaken by the European Commission.

8.4 IMPACT ON NEW SPECTRUM AND WRC-2007

Available frequency bands are the 'real-estate' of mobile and wireless communications for the deployment of new systems and the establishment of new business areas. As seen from

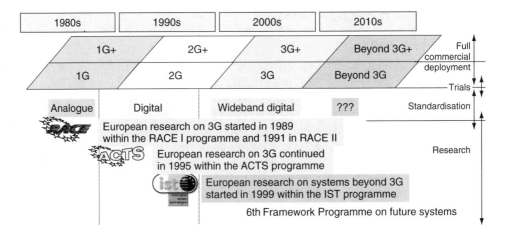

Figure 8.5   The European R&D programmes on mobile communication.

Chapter 5, a sufficient amount of frequency spectrum for certain applications is a prerequisite for the economic success of new services in order to provide sufficient system capacity, quality of service (QoS) to the users at a reasonable cost and to enable competition between providers. In addition, global harmonisation of the use of spectrum is motivated by a strong desire to achieve:

• global terminal mobility and roaming;
• economy of scale of devices and infrastructure equipment;
• reduced implementation complexity.

Global harmonisation of spectrum usage is needed to establish new business areas and to improve the overall business case of new concepts. Research activities have a close link to the identification, allocation and implementation of new frequency spectrum.

Future systems for mobile and wireless applications in the wireless world will be comprised of a variety of different, cooperating radio systems in an heterogeneous environment. Such radio systems are already becoming available, or will shortly be developed for different application areas. They include Body Area Networks (BANs), Personal Area Networks (PANs), Home Area Networks (HANs), Vehicle Area Networks (VANs), systems for short-range communication and WLAN applications in hotspots for low mobility applications, fixed wireless access with low mobility for the interconnection of hotspots and new wideband wide-area systems for high mobility (Figure 8.6).

All of these systems have different requirements on the propagation conditions and the impact on mobility. Therefore, having different frequency ranges for the different application areas may be appropriate. This would allow the allocation of the entire traffic to different frequency ranges and the most efficient use of the frequency spectrum.

New concepts, such as the new radio systems for systems beyond 3G, will require larger carrier and system bandwidth than today's systems, in order to provide significantly higher

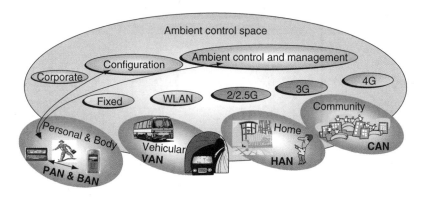

Figure 8.6 Network management composability and reconfigurability. *Source*: Evans (2004).

throughput for the requested QoS. Therefore, new spectrum will be needed. The justification for new spectrum requires work on new methods for spectrum usage, the improvement of the efficiency of radio systems, improved network planning and operation methods, services and applications, traffic engineering models and the derivation of technical requirements based on services and applications requirements. In addition, research is needed on radio propagation for candidate frequency bands for new spectrum.

Following initial research on new concepts for frequency usage, regulators in different regions are working on new methods of flexible frequency usage and frequency sharing methods through adaptive mechanisms and on technology-independent licensing and spectrum allocation schemes. Frequency trading is also on the agenda. There is a trend towards non-dedicated frequency allocations. Such methods could improve the overall frequency usage significantly, if appropriately implemented. Cognitive radio and software defined radio concepts, e.g. (European Commission, n.d.), will play an important role in supporting such new regulatory schemes. Research is needed on the impact of new methods of spectrum usage, e.g. dynamic spectrum allocation (Figure 8.7), its impact on the system design and on coexistence issues between different technologies, to guarantee the friendly cooperation of these different radio technologies, when they are allocated to adjacent bands or operate in frequency sharing mode.

The risks and difficulties resulting from 3G licensing were explained in some detail in Chapter 6 and concerns are being raised in the community concerning the early identification of new frequency spectrum at WRC-2007, the expected licensing process and potentially related high spectrum cost for systems beyond 3G. This may prevent or hinder innovation in Europe compared to other regions. As mentioned earlier, Asia is already very active in research on systems beyond 3G and North America is developing a series of standards in IEEE 802.xx for different application areas.

According to the current ITU-R time schedule new spectrum for systems beyond 3G is intended to be identified at WRC-2007 and should be implemented in the time frame up to and beyond the year 2012. The UMTS Forum's SAG, ERO's ECC PT1 and ITU-R

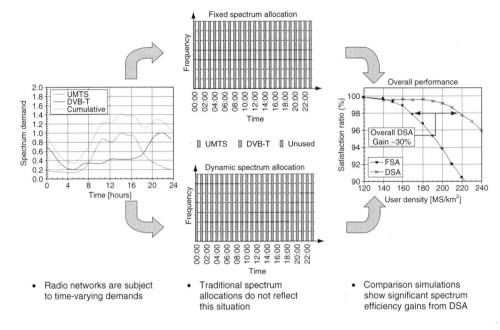

Figure 8.7   Spectrum issues. *Source*: Evans (2004).

WP8F are doing a lot of preparation work on this subject. Europe's 7th Framework Programme will be in a good position to contribute technical results regarding the different aspects of the implementation and most efficient use of the frequency spectrum for systems beyond 3G. Figure 8.8 gives a clue where spectrum for systems beyond 3G may be found.

The term 'mobile broadband system' (MBS) has been used in the past (1990s) in the ITU-R arena when people talked about systems beyond 3G. In some publications it is still used and

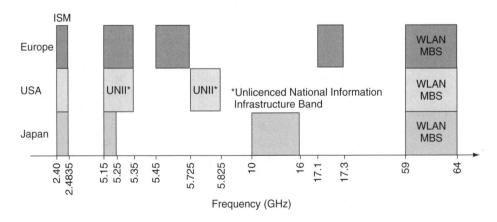

Figure 8.8   New frequency bands for systems beyond 3G? *Source*: UMTS Forum SAG.

means transmission speed up to 100 Mb/s or higher. It is one of many technical components discussed for 4G and reflects some of the research work which is under development in e.g. NTT DoCoMo's laboratories.

The international standardisation after WRC-2007 has to be supported by detailed link- and system-level simulations and larger scale trials. The scope of work should be on the design of the different required radio interface systems using complex signal processing algorithms, including advanced antenna solutions, their commonalities in order to ease interworking, and implementation issues for the development of larger-scale real-time trial systems. In addition, the impact of new frequency ranges after WRC-2007 needs to be investigated. New methods of flexible and adaptable frequency usage and sharing methods (see Figure 8.7) as well as the impact on the system design and implementation need to be considered.

8.5 IMPLICATIONS FOR SYSTEM DESIGN

Bringing the new radio access concept closer to market introduction is in the focus of this research area. Nevertheless, the evolution of the concept also needs to be prepared by pursuing basic research into promising techniques.

Corresponding prototypes have to support the needs of developed and developing country markets in order to provide technologies for the next growth markets for mobile and wireless communication. The outcome of the research should result in a consolidated global consensus approach regarding the defined access technologies.

But not only access technologies will be affected by this research; it will have also a tremendous impact on network design and protocol evolution. Professor Barry Evans, Pro-Vice Chancellor for Research and Enterprise at the University of Surrey, highlighted at DWRC's 5th World Wireless Conference the 'visions of mobile communications beyond 3G'. The last of his figures used at that conference shows how many changes at the protocol level will move wireless communications into its prosperous future (Figure 8.9). The main work items next to the spectrum issue will be

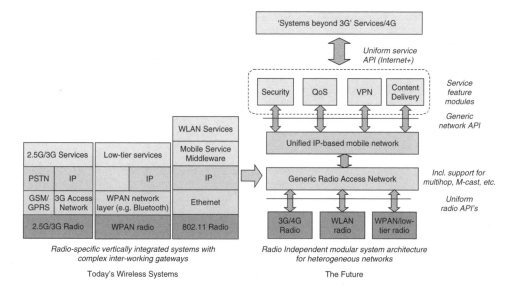

Figure 8.9   'Systems beyond 3G' issues – protocol evolution. *Source*: Evans (2004).

on the generic radio access technology, an important part for 3GPP and ITU-R, a unified IP-based mobile network, something of great relevance for e.g. IETF, and a lot of work on the services front, regarding e.g. QoS, security, content etc., a special task for organisations like OMA, GSMA and the UMTS Forum. In the next few years up to WRC-2007 many steps should be established to make systems on 3G and beyond happen.

References

3G Americas, UMTS to mobilise the data world. http://www.3gamericas.org.

3GPP (2000) Document 3GPP/OP#3(00)43 of the MRP, 3rd Meeting of the Organisation Partners of 3GPP, Beijing, China, 18–19 July. www.3gpp.org.

Alcatel *et. al.* (1998) Working assumptions on TDD and FDD Modes, ETSI STC SMG2, meeting no. 25, Tdoc SMG2 112/98, Geneva, 23–27 February.

Analysys/Intercai (1997) UMTS market forecast study, final report for EC DGXIII, report no. 97043, 12 February.

Analysys Research (2004) European 3G users to begin surge in 2005, April.

CEPT/ERC (1998) Draft report on the introduction of economic criteria in spectrum management and the principles of fees and charging in the CEPT, document RR(97)135.

CNet News and CTIA Daily News (2004a) Camera phones face ban in courts, 7 June 7.

CNet News and CTIA Daily News (2004b) FCC may create more spectrum for wireless broadband, 16 April 16.

Digital World Research Centre (2004) 5th World Wireless Conference, University of Surrey, Guildford, 15–16 July.

EMC (2004a) Mobile subscriber numbers exceed 1.5 billion, 22 June.

EMC (2004b) Mobile subscribers to top 2B by 2006, 22 June.

ERC (1998) Adjacent band compatibility between UMTS and other services in the 2 GHz band, ERC report TG1/02, December.

ERO (1996) ERO report on UMTS, Copenhagen, September.

ETSI (1997a) Concept Alpha Group: Wideband Direct-Sequence CDMA (WCDMA) Evaluation Document (3.0), ETSI SMG#24 Tdoc 905/97.

ETSI (1997b) Delta Concept Group: Enhanced TD-CDMA – A Revolution for UMTS and an Evolution of GSM, ETSI SMG#24 Tdoc 1023/97.

ETSI (1998) Consensus Decision on the UTRA concept to be refined by ETSI SMG2, Tdoc SMG 39/98, SMG#24bis, 28–29 January.

European Commission (1997) Strategy and policy orientations with regard to the further development of mobile and wireless communications (UMTS), COM(97)513, 15 October.

European Commission (1999) The co-ordinated introduction of a third-generation mobile and wireless communications system (UMTS) in the Community, Decision no. 128/1999/EC of the European Parliament and of the European Council.

European Commission (2002) EU Framework Directive 2002/21/EC, 7 March.

European Commission (n.d.) End-to-end reconfigurability, IST Sixth Framework Programme, project reference no. 507995, project coordinator Didier Bourse, Motorola.

Evans B.G. (2004) Managing wireless communications, 5th Wireless World Conference, University of Surrey, Guildford, 15–16 July.

Eylert B. (2002) I get around: 3G and global terminal circulation, 3GSM Show Daily, Cannes, February.

Eylert B. (n.d.) Audiência Pública 198, Manifestação Oral do UMTS Forum, A Utilização das Frequências sob uma Ótica Global.

FierceWireless (2004a) T-Mobile HotSpots generate $1.4M in monthly revenue, 26 March.

FierceWireless (2004b) China's TD-SCDMA fails tests, lags behind other 3G technologies, 9 November.

France Telecom Mobiles (1998) Possible spectrum allocation scenarios to handle UMTS traffic asymmetry, UMTSF SAG, Paris, April.

Herschel Shosteck Associates (1999) Third generation wireless (3g): why, when and how it will happen, November.

Hewitt T. (1998) A discussion document on spectrum asymmetry in the UMTS context, UMTSF SAG, Berlin, March.

Hewitt T. (2000) Report on the IMT-2000 results at WRC-2000, UMTS Forum.

Hillebrand F. (ed.) (2002) *GSM and UMTS*, John Wiley & Sons, Chichester, chapter 15, pp. 385–406.

Huber A.J. and Huber J.F. (2002) *UMTS and Mobile Computing*. Artech House, Boston, MA.

Huber J.F. (1998a) UMTS-spectrum demand per operator – a market oriented study on 'how to deal with traffic asymmetry', UMTSF SAG, Berlin, March.

Huber J.F. (1998b) Spectrum use for low/high mobility radio layers – how can spectrum be optimally applied?, UMTSF SAG, Berlin, March.

IDG News Service and CTIA Daily News (2004) Wi-Fi security and quality improvements coming this fall, 6 May.

IEE Commentary (2004) Mobile video: how long before it takes off? 10 June.

Independent Expert Group on Mobile Phones (IEGMP) (2000) Mobile phones and health ('the Stewart Report'). Available at: www.iegmp.org.uk/report/index.htm.

Interfax (2004) 3G – China's next generation mobile technology, June.

International Telecommunication Union (ITU) (1999) World telecommunications development report 1999, Geneva.

International Telecommunication Union (ITU) (2002a) World telecommunications development report 2002, Geneva, March.

International Telecommunication Union (ITU) (2002b) The Internet for a mobile generation, ITU Internet reports, Geneva, September.

International Telecommunication Union (ITU) (2002c) WP 8A document 8/73-E, 19 June.

ITU-R (1997) Economic aspects of spectrum management, ITU-R Report SM.2012.

ITU-R (1998a) Terrestrial spectrum requirement for IMT-2000, ITU-R TG8-1/120-E, October.

ITU-R (1998b) Working document towards a Preliminary Draft New Report ITU-R M, ITU-R TG8-1/226-E, November.

ITU-R (1998c) ITU-R TG8-1/182, Figures contained in ITU contribution from the United States, 3 November.

ITU-R (1998d) ITU-R TG8-1/184, Figures contained in ITU contribution from Japan, 3 November.

ITU-R (1998e) ITU-R TG8-1/123, Figures contained in ITU TG 8/1 working document, 19 November.

Kyoung Il Kim (ed.) (1999) *Handbook of CDMA System Design, Engineering, and Optimization*. Prentice Hall, Upper Saddle River, NJ.

Lough D.L., Blankenship T.K., Krizman K.J. (n.d.) A short tutorial on wireless LANs and IEEE 802.11, Virginia Polytechnic Institute and State University. Available at: www.computer.org/students/looking/summer97/ieee802. htm.

Lovejoy B. (2003) 3G and Wi-Fi: friend or foe? ITU Telecom World, On-Line News Service.

Merrill Lynch (2000) Wireless Internet – more than voice: the opportunity and issues, 5 June.

Milgrom P. (1989) Auctions and bidding – a primer, *Journal of Economic Prospectives*, no. 3.

Mouritsen R.H. (2002) Telecommunications Act of 1996: relationships to functional theory. *Perspectives: Electronic Journal of the American Association of Behavioral and Social Sciences*, 5. Available at: aabss.org/journal2002/ Mouritsen.htm.

Neue Zürcher Zeitung (NZZ) (2004) Drohendes UMTS-Moratorium, 27 May.

Nieminen T. (1996) Report on www traffic measurements, Mobile ATM Project, Helsinki University of Technology, Communications Laboratory, 29 October.

Ofcom (2004) The communications market 2004, inaugural study of media and telecommunications world, 11 August.

Ovum Research (2004) Trends in mobile data traffic pricing: who to charge? What? When?, 15 June.

RACE (1995) RACE – Mobile Telecommunications Summit, Cascais, 22–24 November.

RCR Wireless (2004) Wireless Internet users to reach 600M by '08, Probe says, 30 April.

Rosston G.L. and Steinberg J.S. (1997) Using market-based spectrum policy to promote the public interest. 50 *Federal Communications Law Journal*, (1), 88–118.

SAG (1998) SAG report, Annex 3: Spectrum Calculations for Terrestrial UMTS Rel. 1.2, 12 March.

SAG (2004) Characteristics of traffic asymmetry relevant to the UMTS frequency arrangement in the 2.6 GHz band, SAG doc 41/22, input document for ECC/PT1 12th Meeting, Helsinki, 19–21 January.

SAG (n.d.) UMTS Forum, SAG doc 41/23.

Schwarz E. (2004) Wi-Fi security standard to require new hardware – 802.11i uses AES encryption. *InfoWorld*, 7 May.

Sidenbladh T. (2002) License and regulatory update on UMTS, ITU Sub-Regional Seminar on IMT-2000, Moscow, 10–13 September.

Siemens (2004) Mobile communication and technology platform, final report, 9 June, Brussels.

Silicon.com (2004) More Wi-Fi hotspots for BT customers?, 28 July.

Silva J. (2004) FCC opens 2.5 GHz spectrum for wireless broadband, *RCS Wireless News*, 10 June 10.

Tafazolli R. (ed.) (2004) *Technologies for the Wireless Future. Wireless World Research Forum (WWRF)*. John Wiley & Sons, Chichester.

UBS Warburg (2000) European ISP/portals, June.

UMTS Forum (1997) A regulatory framework for UMTS, report no. 1, June.

UMTS Forum (1998a) The impact of licence cost levels on the UMTS business case, report no. 3.

UMTS Forum (1998b) Considerations of licensing conditions for UMTS network operations, report no. 4.

UMTS Forum (1998c) Minimum spectrum demand per public terrestrial UMTS operator in the initial phase, report no. 5, February.

UMTS Forum (1998d) UMTS/IMT-2000 spectrum, report no. 6, December.

UMTS Forum (1999a) Report on for UMTS/IMT-2000 terrestrial component, report no. 7, March.

UMTS Forum (1999b) The future mobile market, report no. 8, March.

UMTS Forum (2000a) The UMTS third generation market: structuring the service revenue opportunities, report no. 9, September.

UMTS Forum (2000b) Enabling UMTS/third generation services and applications, report no. 11, October.

UMTS Forum (2001a) Naming, addressing and identification, issues for UMTS, report no. 12, May.

UMTS Forum (2001b) The UMTS third generation market, phase II: structuring the service revenue opportunities', report no. 13, April.

UMTS Forum (2001c) The UMTS third generation market study update, report no. 17, December.

UMTS Forum (2002a) Long term potential remains high for 3G mobile data services, report no. 18, February.

UMTS Forum (2002b) Benefits and drawbacks of introducing a dedicated top level domain within the UMTS environment, report no. 19, February.

UMTS Forum (2002c) Impact and opportunity: public wireless LANs and 3G business revenues, report no. 22, July.

UMTS Forum (2003a) Social shaping of UMTS – preparing the 3G customer, Report of the Digital World Research Centre (DWRC) by J. Vincent and R. Harper on behalf of the UMTS Forum, published by the UMTS Forum as report no. 26, January 2003.

UMTS Forum (2003b) 3G offered traffic characteristics', report no. 33, November.

UMTS Task Force (1996) The road to UMTS – in contact anytime, anywhere, with any one, Brussels, 1 March.

Urban J., Wisely D., Bolinth E., Neureiter G., Liljeberg M., and Robles Valladares T. (2001) BRAIN – an architecture for a broadband radio access network of the next generation. *Wireless Communication and Mobile Computing*, 1(1), 55–75.

Walke B.H. (2000) *Mobile Radio Networks*. John Wiley & Sons, New York.

Walke B., Seidenberg P. and Althoff M.P. (2003) *UMTS – The Fundamentals*. John Wiley & Sons, Chichester.

Wall Street Journal (2004) Vodafone struggles to build global brand in Japan, 16 June.

Wall Street Journal, CTIA Daily News (2004) Globalstar completes bankruptcy restructuring, 15 April.

World Customs Organization (1990) Istanbul Convention, Customs Cooperation Council (CCC), 26 June, entered into force 27 November 1993. UN Office for the Coordination of Humanitarian Affairs, Customs Arrangements to expedite delivery of emergency humanitarian assistance, Geneva.

WTO (2002) WTO Reference Paper on regulatory principles, see www.wto.org/english/tratop e/serve/telecom e/ tel23 e.htm. March (2005).

Yankee Group (2004) Mobile data to provide all revenue growth in W. Europe. *RCR Wireless News*, 17 June.

Zwamborn APM, Vossen SHJA, van Leersum BJAM, Ouwens MA, Mäkel WN (2003) Effects of Global Communication system radio-frequency fields on Well Being and Cognitive Functions of human subjects with and without subjective complaints. *TNO report FEL-03-C148*. www.ez.nl/ (March 2005).

Abbreviations

3G	Third Generation
3GIG	Third Generation Interest Group
3GPP	Third Generation Partnership Project
3GPP2	3GPP's counterpart for the ITU-R IMT-MC standard
AAA	Authorisation, Authentication, Accounting
ACK	Acknowledgment
ACTS	Advanced Communication Technologies and Services
AES	Advanced Security Standard
AH	IP Authentication Header
AMPS	American Mobile Phone System
ANATEL	Brazilian Telecommunications Regulator
ANSI	American National Standards Institute
API	Application Programme Interface
APN	Access Point Names
APNIC	Asia Pacific Network Information Centre
ARIN	American Registry for Internet Numbers
ARPU	Average Revenue Per User
ATM	Asynchronous Transfer Mode
AuC	Authentication Centre
B2B	Business to Business
B2C	Business to Customer
BAN	Body Area Network
BCD	Binary Coded Decimal
BG	Border Gateway
BRAIN	Broadband Radio Access for IP Networks
BSC	Base Station Controller
BTS	Base Transceiver Station
BUWAL	Bundesamt für Umwelt, Wald und Landschaft (Switzerland)
C2C	Consumer to Consumer
CAA	Conformity Assessment Agreement
CAGR	Compound Annual Growth Rate
CBD	Central Business District
CCIR	Comité Consultatif International de Radiocommunication
CCITT	Comité Consultatif International Téléphonique et Télégraphique
CC/PP	Composite Capability/Preference Profile
ccTLD	Country Code Top Level Domain

The Mobile Multimedia Business: Requirements and Solutions   Bernd Eylert
© 2005 John Wiley & Sons, Ltd

CCTV	Closed Circuit Television
CDMA	Code Division Multiplex Access
CEPT	Conférence Européenne des Administrations, des Postes et des Télécommunications
CIDR	Classless Inter-Domain Routing
CITEL	Inter-American Telecommunication Commission
CLI	Calling Line Identity
CPM	Conference Preparatory Meeting
CPU	Central Processing Unit
CRC	Cyclic Redundancy Check
CS	Circuit Switched
CSCF	Call State Control Function
CSMA/CA	Carrier Sense Multiple Access with Collision Avoidance
CT	Cordless Telephone
CTIA	Cellular Telecommunications and Internet Association
CTS	Clear-To-Send
DECT	Digital Enhanced Cordless Telecommunications
DCS	Digital Communication System
DHCP	Dynamic Host Configuration Protocol
DL	Downlink
DNS	Domain Naming Scheme/Domain Name Server
DSL	Digital Subscriber Line
DSSS	Direct Sequence Spread Spectrum
DVB-T	Digital Video Broadcasting – Terrestrial
DWRC	Digital World Research Centre
E2R	End-to-End Reconfigurability
EBPP	Electronic Bill Presentation and Payment
EC	European Commission
ECC	European Communications Committee
EDGE	Enhanced Data Rates for GSM Evolution
EIR	Equipment Identity Register
EITO	European Information Technology Observatory
EMEA	Europe, Middle East, Africa
ERC	European Radiocommunications Committee
ERO	European Radiocommunications Office
ERMES	European Radio Messaging System
ESP	IP Encapsulation Security Payload
ETRI	Electronics and Telecommunications Research Institute, Korea
ETSI	European Telecommunications Standards Institute
EU	European Union
FCC	Federal Communications Commission (the US regulator)
FDD	Frequency Division Duplex
FDMA	Frequency Division Multiplex Access
FER	Failure Error Rate
FOMA	Freedom of Mobile Multimedia Access (Trade name of NTT DoCoMo's 3G system)
FPLMTS	Future Public Land Mobile Telecommunication System
FTP	File Transfer Protocol
GATS	General Agreement on Trade in Services
GDP	Gross Development Product
GEO	Geostationary Orbiter
GERAN	GSM EDGE Radio Access Networks
GGSN	Gateway GPRS Support Node
GLR	Gateway Location Register
GPRS	General Packet Radio Service
GRID	Global Resource Information Database
GSM	Group Spéciale Mobile; Global System for Mobile Communication

GSMA	GSM Association
GSM MoU	GSM Memorandum of Understanding (today the GSM Association)
GSN	GPRS Support Nodes
gTLD	Generic Top Level Domain
GTP	GPRS Tunnelling Protocol
HAN	Home Area Network
HAPS	High Altitude Platform Station
HE	Home Environment
HF	High Frequency
HIPERLAN	HIgh-Performance Local Radio Access Network
HLR	Home Location Register
HomeRF	Home Radio Frequency Standard
HSDPA	High Speed Downlink Packet Application
HTML	Hypertext Markup Language
http	Hypertext Transfer/Transport Protocol
ICANN	Internet Corporation for Assigned Names and Numbers
ICT	Information Communication Technology
IDPR	Internet Detail Protocol Record
IEC	International Electrotechnical Commission
IEEE	Institute of Electrical and Electronics Engineers
IEEE 802.16/802.xx	IEEE standards series
IETF	Internet Engineering Task Force
IKE	Internet Key Exchange
IKMP	Internet Key Management Protocol
ILO	International Labour Organization
IMEI	International Mobile Equipment Identifier
IMO	International Maritime Organization
IMS	IP Multimedia Subsystem
IMSI	International Mobile Station Identifier
IN	Intelligent Network
Inmarsat	International Maritime Satellite Organisation
IP	Internet Protocol
IPDR	Internet Detail Protocol Record
IPv4/IPv6	Internet Protocol version 4, respectively 6
IRG	Inter-Regulatory Group
IRPS	Incremental Revenue Per 3G Subscriber
IPSec	IP Security (a Protocol Task Group of IETF)
ISDN	Integrated Services Digital Network
ISM	Industrial, Scientific and Medical
ISO	International Organization for Standardization
ISP	Internet Service Provider
ISTAG Vision	Information Society Technical Advisory Group Vision
ITU	International Telecommunication Union
ITU-R	International Telecommunication Union – Radiocommunication
IUR	International User Requirements
IVDS	Interactive Video Distribution Systems
LAN	Local Area Network
LBS	Location-Based Services
LEO	Low Earth Orbiter
LIR	Local Internet Registries
MAC	Medium Access Control/Message Authentication Code
MAG	Market Aspects Group
MAP	Mobile Application Part
MATS	Funktelefonsystem A
MBMS	Mobile Broadcast and Multicast Service

MBS	Mobile Broadband System
MCC	Mobile Country Code
ME	Mobile Equipment
MEO	Medium Earth Orbiter
MIP	Medium Interface Point
MIPv6	Mobile IP version 6
MMI	Man–Machine Interface
MMP	Mobile Multimedia Portal Platform
MMS	Multimedia Messaging Service
MNC	Mobile Network Code
MoU	Memorandum of Understanding
MPEG	Motion Pictures Expert Group
MPLS	Multi-Protocol Label Switching
MPT	Ministry of Post and Telecommunications
MRP	Marketing Representation Partners
MS	Mobile Station
MSC	Mobile Switching Centre
MSIN	Mobile Station Identification Number
MSISDN	Mobile Subscriber Integrated Services Digital Network
MSRN	Mobile Station Roaming Numbers
MSS	Mobile Satellite Service
M-TLD	Mobile Top Level Domain
MVNO	Mobile Virtual Network Operators
NAT	Network Address Translators
NGBT	Negotiating Group on Basic Telecommunications
NLA	Next Level Aggregator
NMSI	National Mobile Station Identifier
NMT	Nordic Mobile Telecommunications
NRA	National Regulatory Authorities
NSI	Network Solutions Inc.
NZZ	Neue Zürcher Zeitung
Ofcom	Office for Communications (the UK communications regulator)
OFDMA	Orthogonal Frequency Division Multiplex Access
OMA	Open Mobile Alliance
PABX	Private Automatic Branch Exchange
PAN	Personal Area Network
PC	Personal Computer
PCIA	Personal Communications Industry Association
PCMCIA	Personal Computer Memory Card International Association
PDA	Personal Digital Assistant
PDC	Pacific Digital Cellular
PDP	Packet Data Protocol
PHS	Personal Handy Phone System
PKI	Public Key Infrastructure
PLMN	Public Land Mobile Network
PMR	Professional Mobile Radio
PoP	Point of Presence
POCSAG	Post Office Code Standarization Advisory Group
PS	Packet Switched
PSTN	Public Switched Telephone Network
PTO	Public Telecommunications Operators
PTT	Post, Telephone and Telegraph Administration
QoS	Quality of Services
RACE	Research in Advanced Communication Technologies in Europe
R&D	Research and Development

RAG	Regulatory Aspects Group
RAN	Radio Access Network
RFC	Request for Comments
RFID	Radio Frequency Identification
RIPE	Réseaux IP Européens
RIR	Regional Internet Registries
RNC	Radio Network Controller
RoW	Rest of World
RSVP	Resource Reservation Protocol
SAG	Spectrums Aspects Group
SDoC	Supplier's Declaration of Conformity
SDR	Software Defined Radio
SEA	Spokesman Election Algorithm
SGSN	Serving GPRS Support Node
SIM	Subscriber Identity Module
SIP	Session Initiation Protocol
SLA	Site Level Aggregators
SLD	Second Level Domain names
SLP	Service Logic Programme
SMDS	Switch Multimegabit Data Service
SME	Small and Medium Sized Enterprises
SMS	Short Message Service
SOHO	Small Office/Home Office
SWAP	Shared Wireless Access Protocol
TABD	Trans-Atlantic Business Dialogue
TACS	Total Access Communications Service
TETRA	Terrestrial Trunked Radio
TAP	Transfer Account Procedure
TC/IWF	Transcoder and Interworking Function
TD-(S)CDMA	Time Division (Small-band) CDMA
TDD	Time Division Duplex
TDMA	Time Division Multiplex Access
TFTS	Terrestrial Flight Telephone System
TIMES	Telecommunications, Information Technology (IT), Media, Entertainment, Security
TID	Tunnel Identifier
TLA	Top Level Aggregator
TLD	Top Level Domain
TR	Technical Regulation
UAPROF	User Agent Profile
UMTS	Universal Mobile Telecommunication System
UHF	Ultra High Frequency
UI	User Interface
UL	Uplink
UPT	Universal Personal Telecommunications
URL	Uniform Resource Locator
USECA	UMTS Security Architecture
USIM	UMTS Subscriber Identity Module
UTRA	Universal Terrestrial Radio Access
UTRAN	UMTS or Universal Terrestrial Radio Access Networks
UWB	Ultra Wide Band
VAN	Vehicle Area Networks
VHE	Virtual Home Environment
VHF	Very High Frequency
VLR	Visitor Location Register
VMSC	Visiting MSC

VoIP	Voice over IP
VPN	Virtual Private Network
W3C	World Wide Web Consortium
WAN	Wide Area Network
WAP	Wireless Access Protocol
WARC	World Administrative Radiocommunications Conference
WASP	Wireless Application Service Provider
WCS	Wireless Communications Services
WEP	Wired Equivalent Privacy
WiFi/Wi-Fi	Wireless Fidelity (American synonym for WLAN)
WIPO	World Intellectual Property Organization
WLAN	Wireless Local Area Network
WME	Wireless Media Extensions
WML	Wireless Markup Language
WPA	Wi-Fi Protected Access
WRC	World Radiocommunication Congress
WSDL	Web Service Description Language
WTO	World Trade Organization
WWRF	Wireless World Research Forum
XML	Extensible Markup Language

Index